SpringerBriefs in Optimization

Series Editors

Panos M. Pardalos
János D. Pintér
Stephen M. Robinson
Tamás Terlaky
My T. Thai

SpringerBriefs in Optimization showcases algorithmic and theoretical techniques, case studies, and applications within the broad-based field of optimization. Manuscripts related to the ever-growing applications of optimization in applied mathematics, engineering, medicine, economics, and other applied sciences are encouraged.

For further volumes:
http://www.springer.com/series/8918

Stefan Scholtes

Introduction to Piecewise Differentiable Equations

 Springer

Stefan Scholtes
Department of Engineering
University of Cambridge
Cambridge, UK

ISSN 2190-8354   ISSN 2191-575X (electronic)
ISBN 978-1-4614-4339-1 ISBN 978-1-4614-4340-7 (eBook)
DOI 10.1007/978-1-4614-4340-7
Springer New York Heidelberg Dordrecht London

Library of Congress Control Number: 2012942479

Mathematics Subject Classification (2010): 90C26, 90C31, 90C33 and also 80M50, 34A34

Printed on acid-free paper

Springer is part of Springer Science+Business Media (www.springer.com)

Introduction to Piecewise
  Differentiable Equations
  Stefan Scholtes

Preface

The aim of this manuscript is to provide an introduction to the theory of piecewise differentiable functions and, specifically, piecewise differentiable equations. The presentation is based on two basic tools for the analysis of piecewise differentiable functions: the Bouligand derivative as the analytic approximation concept and the theory of piecewise affine functions as the combinatorial tool for the study of the approximation function.

The first chapter presents two sample problems which illustrate the relevance of the study of piecewise differentiable equations. Chapter 2 then investigates piecewise affine functions and piecewise affine equations, followed by an introduction to the Bouligand derivative in the third chapter. Chapter 4 is concerned with piecewise differentiable functions and combines the results of the former chapters to develop inverse and implicit function theorems for piecewise differentiable equations. The final chapter presents two applications of the results to equilibrium modeling and parametric optimization.

This booklet is a reprint of my Habilitation Thesis of 1994. Although the history of piecewise differentiable equations can be traced back to J.H.C. Whitehead, the theory had not been uniformly presented when I wrote my thesis. In particular, its relation to more recent developments in nonsmooth analysis and the theory of piecewise affine mappings had not been sufficiently integrated. The manuscript of the thesis has remained in sufficient demand to warrant its publication in this series. In order to make the theory accessible to a large audience, I have tried to keep the mathematical prerequisites at a minimal level. In fact, most parts of this booklet can be understood with a basic knowledge of mathematical analysis. The treatment of piecewise affine functions requires some familiarity with polyhedral theory at the level of a standard course in linear programming. I hope the text will serve graduate and advanced undergraduate students as a gentle introduction to the theory of piecewise smooth functions and equations in finite dimensions and equip them with basic mathematical intuition on the combinatorial aspects of nonsmooth analysis that will help them conquer more advanced and more recent material. I am grateful to Steve Robinson for his encouragement to publish the manuscript in this series.

Acknowledgements I am grateful for the help of my colleagues Sven Bartels, Bert Jongen, Diethard Klatte, Bernd Kummer, Ludwig Kuntz, Klaus Neumann, Diethard Pallaschke, Peter Recht, Danny Ralph, Steve Robinson, and Wolfgang Weil. All errors remain my own.

Cambridge   Stefan Scholtes

Contents

Chapter 1
Sample Problems for Nonsmooth Equations

To illustrate the latter idea, suppose we are interested in a vector which is known to minimize a function f subject to equality constraints $h(x) = 0$, where $f : \mathbb{R}^n \to \mathbb{R}$ and $h : \mathbb{R}^n \to \mathbb{R}^m$ are twice differentiable. Instead of working with the realistic minimization model, one may try to determine the unknown quantity by specifying a set of necessary optimality conditions. Assuming that $\nabla h(x)$ has full row rank at every $x \in \mathbb{R}^n$ with $h(x) = 0$, a necessary condition for a point x to be a minimizer of f subject to $h(x) = 0$ is the existence of a vector $\lambda \in \mathbb{R}^m$ such that the pair (x, λ) satisfies the stationary point conditions

$$\nabla f(x) + \nabla h(x)^T \lambda = 0,$$
$$h(x) = 0. \tag{1.1}$$

Note that a reasonable set of necessary conditions should not leave us with too many candidates for the demanded quantity. If, for instance, the conditions are formulated as a system of linear or differentiable equations, one would like to have as many equations as there are unknown quantities, so that chances are good to have a unique or at least locally unique solution. This is indeed the case for the stationary point conditions (1.1). What are the advantages of passing from the original minimization problem to the stationary point conditions (1.1)?

Existence and uniqueness questions: The theory of differentiable equations provides us with a variety of tools to check whether a solution exists and whether it is unique. For instance if the C^1-function $F : \mathbb{R}^n \to \mathbb{R}^n$ is a diffeomorphism, i.e., if it has a differentiable inverse function, then the equation $F(x) = 0$ certainly has a unique solution. A necessary and sufficient condition for the function F to be a diffeomorphism is the nonsingularity of the Jacobian $\nabla F(x)$ at every point $x \in \mathbb{R}^n$ and the closeness of the function F, i.e., the images $F(A)$ of closed sets $A \subseteq \mathbb{R}^n$ are closed. A sufficient condition is provided by Hadamard's theorem which states that F is a diffeomorphism if the Jacobians $\nabla F(x)$ are nonsingular and there exists a constant $\alpha \in \mathbb{R}$ such that $\|\nabla F(x)^{-1}\| \leq \alpha$ for every $x \in \mathbb{R}^n$.

S. Scholtes, *Introduction to Piecewise Differentiable Equations*, SpringerBriefs in Optimization, DOI 10.1007/978-1-4614-4340-7_1, © Stefan Scholtes 2012

Algorithms: We can attempt to solve the problem by applying Newton's methods or one of its variants to the system of equations. If this method yields a solution (x^*, λ^*) of the equation, then the local optimality of x^* for the minimization problem can be verified by the use of a second-order sufficiency condition like the positive definiteness of the matrix

$$\nabla^2 f(x^*) + \sum_{i=1}^{m} \lambda_i^* \nabla^2 h(x^*)$$

on the nullspace of the matrix $\nabla h(x^*)$.

Sensitivity analysis: If some of the data of the original problem is uncertain or can be controlled, then it is often reasonable to embed the problem into a parametric family of problems, the parameters reflecting uncertain or controllable quantities. In the case of the stationary point conditions (1.1), we thus obtain a parametric equation $F(x, y) = 0$. Given a solution x^* corresponding to a parameter vector y^*, the following questions naturally arise:

1. Is the solution x^* corresponding to y^* locally unique?
2. If so, is there still a unique solution close to x^* if the parameter is slightly perturbed?
3. If so, what are the continuity and differentiability properties of the locally defined solution function $x(y)$?
4. Can we estimate the asymptotic condition number

$$\limsup_{y \to y^*} \frac{\|x(y) - x(y^*)\|}{\|y - y^*\|}$$

of the solution function $x(y)$ at the point y^*?

In the differentiable case, these questions are all answered by the implicit function theorem which states that the nonsingularity of the reduced Jacobian $\nabla_x F(x^*, y^*)$ implies the existence of a locally unique solution function $x(y)$ of the equation $F(x, y) = 0$ which, in addition, is continuously differentiable with

$$\nabla x(y^*) = -\nabla_x F(x^*, y^*)^{-1} \nabla_y F(x^*, y^*).$$

In particular, the norm of the latter matrix is the asymptotic condition number of the function $x(y)$ at y^*.

If a problem is more complicated than the equality-constrained minimization problem, then necessary conditions often naturally involve inequalities in addition to the equations. However, the trivial observation that $f(x) \leq 0$ if and only if $\max\{f(x), 0\} = 0$ shows that inequalities can be readily turned into equations if the differentiability assumption is dropped. The aim of this book is to explain how the above ideas can be generalized to equations which are not necessarily continuously differentiable. We begin with a presentation of two sample problems which can be naturally reformulated in terms of nonsmooth equations. To illustrate the practical relevance of the sample problems, we first introduce a practical application, the mathematical formulation of which leads to the general problem.

1.1 Complementarity Problems

1.1.1 Equilibria of Dynamical Systems

The theory of dynamical systems is to a large extent concerned with the analysis of equilibrium states. We will not go into the details of this theory but instead use as a working definition for a dynamical system a tuple (X, T), where X denotes a set of admissible states of the system, called the state space, while T is a set-valued transition mapping which assigns to a given state $x \in X$ a set $T(x) \subseteq X$ of admissible subsequent states. If the system is in state x, we assume that there is an incentive for the system to move from state x to some state $y \in T(x)$. We neither assume knowledge about the time when the change occurs nor about the actual choice of the state $y \in T(x)$. The system is said to be in an equilibrium state if there is no incentive to change the state, i.e., $x \in X$ is an *equilibrium state* of the system (X, T) if $T(x) = \emptyset$.

An elementary physical illustration of a dynamical system is a particle moving in space and being driven by a vector field $f : \mathbb{R}^3 \to \mathbb{R}^3$. The position $x(t)$ of the particle at time t changes according to the differential equation

$$\frac{dx}{dt} = f(x).$$

In our terminology, $X = \mathbb{R}^3$ is the state space of the system, while the transition mapping T may be defined by

$$T(x) = \left\{ y \in \mathbb{R}^3 \mid f(x)^T y > f(x)^T x \right\}.$$

The equilibrium condition $T(x) = \emptyset$ corresponds to the well-known condition $f(x) = 0$. If the particle is bound to a convex subset X of $\mathbb{R}^3$, then the transition mapping T is naturally defined by

$$T(x) = \left\{ y \in X \mid f(x)^T y > f(x)^T x \right\},$$

and the equilibrium condition $T(x) = \emptyset$ is equivalent to the variational inequality

$$f(x)^T x \geq f(x)^T y \quad \text{for every } y \in X. \tag{1.2}$$

Beside the classical physical interpretations, there are interesting economical situations which can be put into the framework of a dynamical systems. Consider for instance a closed market with n firms, each firm producing a single product. Suppose c_i is the unit production cost of firm i, while p_i is the price per unit of the commodity produced by firm i. The production costs may depend on the output x_i of the ith firm, i.e., $c_i = c_i(x_i)$, while, due to substitution effects, the price p_i of the commodity produced by the ith firm may also depend on the outputs of all other firms, i.e., $p_i = p_i(x_1, \ldots, x_n)$. Defining the function $d_i(x) = p_i(x) - c_i(x_i)$, the total profit of the ith firm is given by $d_i(x)x_i$. The aim of each firm manager is to maximize total profits. Observing a vector of joint outputs $x = (x_1, \ldots, x_n)$, the firm managers may increase or decrease the output of their firms. We assume that the manager of the ith firm knows the cost function c_i and that he has an idea how the price p_i changes if he changes the output of his firm, provided the outputs of all other firms remain constant. The manager of the ith firm will increase the output of his firm by some amount $\Delta > 0$ if

$$d_i(x + \Delta e_i)(x_i + \Delta) \geq d_i(x)x_i,$$

where e_i denotes the ith unit vector in $\mathbb{R}^n$. The latter inequality holds if and only if

$$\frac{d_i(x + \Delta e_i) - d_i(x)}{\Delta} x_i \geq -d_i(x + \Delta e_i).$$

Letting Δ tend to zero and assuming the differentiability of the function d_i we thus see that a small increase of the variable x_i results in an increase in total profits as long as

$$\frac{\partial d_i}{\partial x_i}(x)x_i + d_i(x) > 0.$$

Similarly a small decrease of the variable x_i yields an increase in total profits if

$$\frac{\partial d_i}{\partial x_i}(x)x_i + d_i(x) < 0.$$

So we may think of the system as being driven by the vector field $F : \mathbb{R}^n \to \mathbb{R}^n$ defined by

$$F_i(x) = \frac{\partial d_i}{\partial x_i}(x)x_i + d_i(x), i = 1, \ldots, n,$$

The set of admissible states is the nonnegative orthant $X = \mathbb{R}^n_+$. As before, we use the transition mapping

$$T(x) = \{y \in X \,|\, F(x)^T y > F(x)^T x\},$$

and thus a state $x \in \mathbb{R}^n_+$ is an equilibrium state if it satisfies the variational inequality

$$F(x)^T x \geq F(x)^T y \text{ for every } y \in \mathbb{R}^n_+. \tag{1.3}$$

It is not difficult to check that this is the case if and only if

$$x \geq 0,$$
$$F(x) \leq 0,$$
$$F(x)^T x = 0. \tag{1.4}$$

An economic interpretation of the latter conditions is straightforward. In fact, in an equilibrium situation we observe that either the manager of the ith firm has no incentive to change the output x_i, or the ith firm is out of business and there is no incentive to start the production again. In the first situation we thus have $x_i \geq 0$ and $F_i(x) = \frac{\partial d_i}{\partial x_i}(x)x_i + d_i(x) = 0$, while the second situation corresponds to $x_i = 0$ and $F_i(x) = \frac{\partial d_i}{\partial x_i}(x)x_i + d_i(x) \leq 0$. The latter situation occurs if the price $p_i(x)$ of the ith product at the current overall production level x does not exceed the unit production cost $c_i(0)$ of the ith firm at the zero production level.

1.1.2 Nonlinear Complementarity Problems

Problem (1.4) is directly transformed into the *nonlinear complementarity problem* corresponding to a function $f : \mathbb{R}^n \to \mathbb{R}^n$ which is the problem of finding a vector $x \in \mathbb{R}^n$ such that

$$x \geq 0,$$
$$f(x) \geq 0,$$
$$f(x)^T x = 0. \tag{1.5}$$

The nonlinear complementarity problem can be easily transformed into a nonsmooth equation. In fact, $x \in \mathbb{R}^n$ is a solution of (1.5) if and only if

$$\min\{f_i(x), x_i\} = 0, i = 1, \ldots, n.$$

We will see later on that a slightly more complicated formulation has some advantages. This formulation is obtained by introducing an artificial variable $z = f(x)$. One readily verifies that x solves (1.5) if and only if there exists a vector $z \in \mathbb{R}^n$ such that

$$H(x,z) = \begin{pmatrix} f(x) - z \\ \min\{x_1, z_1\} \\ \vdots \\ \min\{x_n, z_n\} \end{pmatrix} = \begin{pmatrix} 0 \\ 0 \\ \vdots \\ 0 \end{pmatrix}.$$

The advantage of the latter formulation will become apparent only later in the text in connection with strongly B-differentiable functions. In fact, if f is continuously differentiable, then the function H is strongly B-differentiable, while the function $\min\{f_i(x), x_i\}$ might not have this property.

1.1.3 Comments and References

For an introduction to the variational inequality problem and its relation to dynamical systems we refer to the recent article [15] of Dupuis and Nagurney. Complementarity theory has been initiated by Lemke and Howson in their article [43] which deals with equilibria of bimatrix games. Starting with the latter application in game theory, complementarity formulations have been used to model a variety of equilibrium situations in economics and engineering. There is still a vivid research activity in the field. For a collection of recent results we refer the interested reader to the publications [10, 11], in particular, to the survey article [25] of Harker and Pang. For a recent contribution dealing with the relation between nonlinear complementarity theory and nonlinear programming we refer to the paper [46] of Mangasarian and Solodov and the references herein.

1.2 Stationary Solutions of Parametric Programs

1.2.1 Multiobjective Optimization

An important problem of the engineering and management sciences is the determination of an optimal decision out of a whole set of feasible decisions. The decision-finding process can often be supported by the use of a mathematical model. Assuming that the different decisions can be identified with finite-dimensional vectors, such a model may be specified by a collection of constraint functions which determine the set of feasible decisions and a set of objective functions which reflect the decision maker's preferences. The aim is to find a feasible decision which simultaneously minimizes the objective functions. Such problems belong to the realm of multiobjective optimization, a discipline which studies various specifications of the term "simultaneous minimization." The most natural solution concept is that of a Pareto-optimal decision which is a feasible decision with the property that there is no other decision which increases at least one of the objectives without decreasing any of the other objectives. Being theoretically very appealing, most solution concepts for multiobjective optimization problems share a severe practical drawback: They do not determine a unique solution for the decision problems, so that the decision maker needs to select his "optimal" decision from a whole set of generated solutions. In particular, it is usually not possible to provide the whole solution set, let alone the problem of representing it in a way which allows the decision maker to find the "optimal" decision. In an approach to solve this problem, one may focus attention on a reasonably large subset of "optimal" solutions and trace this subset with the aid of finitely many parameters. The hope is that the decision maker can identify his optimal parameters in an interactive process. The most natural parameters for multiobjective optimization problems are weights for the objective functions. Depending on such weights, one transforms the multiobjective optimization problem into a parametric mathematical program by minimizing the sum of the weighted objectives over the set of feasible decisions. Such mathematical programs are more likely to have unique optimal solutions. One easily verifies that every solution of a mathematical program corresponding to a strictly positive weight vector is a Pareto-optimal solution of the multiobjective optimization problem. The iterative procedure to determine the decision maker's "optimal" decision is the following. Starting with an arbitrary weight vector, one calculates the corresponding solution of the parametric program and suggests it to the decision maker. The decision maker may either accept the solution or, if he is unhappy with the value of one or another of the objectives, he may increase the corresponding weights in the hope to obtain a better decision. To decide whether the suggested solution is acceptable, the decision maker may ask for

- A local approximation of the set of Pareto-optimal solutions

If the decision maker is unhappy with the value of a specific objective, then he may want to increase the weight of this objective. To control the changes of the other objectives, he may thus be interested in

- The change of the value of the objectives if the weight vector is changed

In order to get an idea how the questions of the decision maker can be answered, we formalize the above approach. We suppose that $M \subseteq \mathbb{R}^n$ is the set of feasible decisions, and that the function $f = (f_1, \ldots, f_p) : \mathbb{R}^n \to \mathbb{R}^p$ reflects the different objectives $f_1, \ldots, f_p$ of the decision maker. Instead of considering Pareto-optimal solutions we confine ourselves to the subset of all solutions of the parametric optimization problem

$$\min \left\{ f(x)^T y \,|\, x \in M \right\}, \tag{1.6}$$

corresponding to strictly positive weight parameters $y \in \mathbb{R}^p$, i.e., we focus attention to the set

$$W = \left\{ z \in M \,|\, z = \mathrm{argmin}_{x \in M} \, f(x)^T y, \, y_i > 0, \sum_{i=1}^{p} y_i = 1 \right\}. \tag{1.7}$$

Let x^* be a unique optimal solution of problem (1.6) corresponding to the parameter vector y^*. Suppose for the moment that we are able to formulate problem (1.6) as the solution of an equation $F(x, y) = 0$ and that the function $F : \mathbb{R}^n \times \mathbb{R}^p \to \mathbb{R}^n$ is continuously differentiable at the point (x^*, y^*). In this case, we can apply the implicit function theorem to answer both questions of the decision maker. In fact, if the assumptions of the latter theorem are satisfied, i.e., if the reduced Jacobian $\nabla_x F(x^*, y^*)$ is nonsingular, then there exist open neighborhoods U of x^* and V of y^* such that for every $y \in V$ the equation $F(x, y) = 0$ has a unique solution $x(y)$ in U. Moreover, the solution function $x(y)$ is a continuously differentiable in V and its Jacobian at the point y^* is given by

$$\nabla x(y^*) = -\nabla_x F(x^*, y^*)^{-1} \nabla_y F(x^*, y^*).$$

In particular, the set

$$W'(x^*) = \left\{ x(y^*) + \nabla x(y^*)(y - y^*) \mid y_i > 0, \sum_{i=1}^{p} y_i = 1 \right\}$$

is a local approximation of the set W defined in (1.7) at the point x^*. If the decision maker is unhappy with the value of the objective f_i, he may be interested in the quantity

$$\nabla f_i(x^*(y^*))^T \frac{\partial x}{\partial y_i}(y^*)$$

which approximates the change of f_i corresponding to a unit increase of the weight y_i. More generally, the decision maker may provide a change vector for the weights, i.e., a vector $v \in \mathbb{R}^p$, the components of which sum up to zero, and be interested in

$$\nabla f_i(x^*(y^*)) \nabla x(y^*) v,$$

a quantity which approximates the change $f_i(y + v) - f_i(y)$ of the objective f_i corresponding to small vectors v.

In the sequel we introduce a formulation of the stationary point conditions for parametric programs as a parametric nonsmooth equation which will eventually allow us to answer the questions of the decision maker analogously to the latter approach for differentiable equations.

1.2.2 The Kojima Mapping

Let us first briefly recall some basic notions and results from mathematical programming. We consider the parametric nonlinear program

$$\mathbf{P(y)} \min_{x \in \mathbb{R}^n} \{ f(x, y) \mid g(x, y) \leq 0, h(x, y) = 0 \},$$

where the functions $f : \mathbb{R}^n \times \mathbb{R}^p \to \mathbb{R}$, $g : \mathbb{R}^n \times \mathbb{R}^p \to \mathbb{R}^l$, and $h : \mathbb{R}^n \times \mathbb{R}^p \to \mathbb{R}^m$ are twice differentiable. Suppose for the moment that the parameter y is fixed.

A general scheme for the development of necessary optimality conditions for a vector x to be a solution of a nonlinear problem is to replace the problem functions by their first-order approximations at the point x and to find conditions which ensure that x is the solution of the resulting linear problem whenever x is the solution of the original nonlinear problem. In the case of the mathematical program $(P(y))$ the vector x is a solution of this approximated problem if and only if the origin is a solution of the linear program

$$\textbf{LP(x,y)} \min_{v \in \mathbb{R}^n} \{\nabla_x f(x, y)v | \nabla_x g(x, y)v \le -g(x, y),$$

$$\nabla_x h(x, y)v = 0\}. \tag{1.8}$$

Duality theory for linear programming shows that this is true if and only if there exist vectors $\lambda \in \mathbb{R}^l$, and $\mu \in \mathbb{R}^m$ such that

$$\nabla_x f(x, y) + \nabla_x g(x, y)\lambda + \nabla_x h(x, y)\mu = 0,$$

$$g(x, y) \le 0,$$

$$h(x, y) = 0,$$

$$\lambda \ge 0,$$

$$g(x, y)^T \lambda = 0. \tag{1.9}$$

The latter conditions are well known as the stationary point conditions or Karush-Kuhn-Tucker conditions for nonlinear programs, i.e., a point $x \in \mathbb{R}^n$ is called a *stationary point* of $(P(y))$ if and only if there exist multiplier vectors $\lambda \in \mathbb{R}^l, \mu \in \mathbb{R}^m$, such that (x, λ, μ, y) satisfy the conditions (1.9). Conditions which ensure that a local minimizer is indeed a stationary point of the problem are called constraint qualifications. For a treatment of such conditions we refer to the standard literature on nonlinear programming.

There are various formulations of the stationary point conditions (1.9) as a system of nonsmooth equation. We follow an approach of M. Kojima, who introduced a mapping $F : \mathbb{R}^n \times \mathbb{R}^{l+m} \times \mathbb{R}^p \to \mathbb{R}^n \times \mathbb{R}^{l+m}$ defined by

$$F(x, v, y) = \begin{pmatrix} \nabla_x f(x, y) + \sum_{i=1}^{l} \max\{v_i, 0\}\nabla_x g_i(x, y) + \sum_{j=1}^{m} v_{l+j}\nabla_x h_j(x, y) \\ -g_1(x, y) + \min\{v_1, 0\} \\ \vdots \\ -g_l(x, y) + \min\{v_l, 0\} \\ h_1(x, y) \\ \vdots \\ h_m(x, y) \end{pmatrix}.$$

One readily verifies that there exist vectors $\lambda \in \mathbb{R}^l$ and $\mu \in \mathbb{R}^m$ such that (x, λ, μ, y) satisfies the stationary point conditions (1.9) if and only if there exists a vector $\nu \in \mathbb{R}^{l+m}$ such that

$$F(x, \nu, y) = 0. \tag{1.10}$$

We call the mapping F the *Kojima mapping* corresponding to the parametric program $(P(y))$. Equation (1.10) thus yields a reformulation of the stationary point conditions for the parametric program $(P(y))$ as a nonsmooth equation.

1.2.3 Comments and References

Many specialists in nonlinear programming have published excellent textbooks on the subject. We confine ourselves to a reference to the classical text [19] of Fletcher, and to Mangasarian's book [45] which contains an extensive treatment of constraint qualifications for nonlinear programs. The Kojima mapping was introduced by Kojima in [31] to investigate sensitivity and stability questions for nonlinear programs.

1.3 Appendix: Differentiability Versus Nondifferentiability

It is well known that every closed subset of $\mathbb{R}^n$ is the solution set of a C^∞-equation. Hence every problem with a closed solution set can, in principle, be formulated as the solution of a C^∞-equation. However, the bare formulation as a differentiable equation does not provide any benefits. To see this, let us take a closer look at a differentiable equation which reformulates the stationary point conditions (1.10) of the parametric program $(P(y))$ introduced in the latter section. Consider the function $G : \mathbb{R}^n \times \mathbb{R}^{l+m} \times \mathbb{R}^p \to \mathbb{R}^n \times \mathbb{R}^{l+m}$ defined by

$$G(x, \alpha, y) = \begin{pmatrix} \nabla_x f(x, y) + \sum_{i=1}^{l} \max\{0, \alpha_i^3\} \nabla_x g_i(x, y) + \sum_{j=1}^{m} \alpha_{l+j} \nabla_x h_j(x, y) \\ -g_1(x, y) + \min\{0, \alpha_1^3\} \\ \vdots \\ -g_l(x, y) + \min\{0, \alpha_l^3\} \\ -h_1(x, y) \\ \vdots \\ -h_m(x, y) \end{pmatrix}.$$

$$\tag{1.11}$$

Clearly G is continuously differentiable. Setting $\alpha_i = \sqrt[3]{v_i}$ for $i = 1,\ldots,l$ and $\alpha_j = v_j$ for $j = l+1,\ldots,l+m$, one readily verifies that the vector (x,v,y) is a zero of the Kojima mapping F if and only if the vector (x,α,y) satisfies the equation

$$G(x,\alpha,y) = 0. \tag{1.12}$$

The reduced Jacobian $\nabla_{(x,\alpha)}G(x,\alpha,y)$ has an $(n+l+m) \times (n+l+m)$ block structure of the form

$$\begin{pmatrix} A & B & C \\ -D^T & E & 0 \\ -C^T & 0 & 0 \end{pmatrix},$$

where

$$A = \nabla_x^2 f(x,y) + \sum_{i=1}^{l} \max\{0,\alpha_i^3\}\nabla_x^2 g_i(x,y) + \sum_{j=1}^{m} \alpha_{l+j}\nabla_x^2 h_j(x,y),$$

$$B = \left(\alpha_1 \max\{0,3\alpha_1\}\nabla_x g_1(x,y),\ldots,\alpha_l \max\{0,3\alpha_l\}\nabla_x g_l(x,y)\right),$$

$$C = \nabla_x h(x,y),$$

$$D = \nabla_x g(x,y),$$

and E is an $l \times l$ diagonal matrix with the diagonal entries

$$\alpha_1 \min\{0,3\alpha_1\},\ldots,\alpha_l \min\{0,3\alpha_l\}.$$

It is not difficult to verify that this matrix is nonsingular if and only if the following conditions are satisfied:

1. None of the multipliers α_i vanishes.
2. The set of all active gradient vectors consisting of the gradients $\nabla_x g_i(x,y)$ with $g_i(x,y) = 0$ together with the gradients $\nabla_x h_j(x,y)$, $j = 1,\ldots,m$, constitute a linearly independent set of vectors.
3. The matrix $V^T A V$ is nonsingular, where the columns of V form a basis of the orthogonal complement of the linear subspace spanned by the set of active gradient vectors.

Note that the first requirement already implies that the Kojima mapping is differentiable at the point (x,v,y), where $v_i = \alpha_i^3$ for $i = 1,\ldots,l$, and $v_j = \alpha_j$ for $j = l+1,\ldots,m$. So nothing is really won in passing from the nondifferentiable equation (1.10) to the differentiable equation (1.12). The main disadvantage of the differentiable equation (1.12) is the fact that it does not account for the inherent nonsmooth structure of the problem. This inherent nonsmoothness is easily visualized by considering the elementary example

$$\min\{-y_1 x_1 - y_2 x_2 | x_1 \geq 0, x_2 \geq x_1, x_1^2 + x_2^2 \leq 1\}.$$

Clearly for positive weights $y = (y_1, y_2)$, the solution is given by

$$x(y) = \begin{cases} \frac{1}{\sqrt{y_1^2 + y_2^2}}(y_1, y_2) & \text{if } y_2 \geq y_1 > 0, \\ \frac{1}{\sqrt{2}}(1, 1) & \text{if } y_1 \geq y_2 > 0. \end{cases}$$

Although the solution is unique for every positive weight vector y, the function $x(y)$ is not differentiable if $y_1 = y_2$. Hence the assumptions of the implicit function theorem cannot be satisfied at these points independently of the formulation of the problem as a differentiable equation. However, the latter points are distinguished by the drastic change of the solution set and thus a local analysis of the function $x(y)$ at these points would reveal valuable information about the behavior of the solution function. It turns out that it is indeed possible to gather information about the behavior of the solution function at such points if the problem is reformulated as a nonsmooth equation with the aid of the Kojima mapping. However, such information can only be obtained at the cost of a more sophisticated analysis since the classical implicit function theorem will not work. A presentation of the necessary mathematical background is the main subject of this book.

Chapter 2
Piecewise Affine Functions

2.1 Elements from Polyhedral Theory

We begin this chapter with a review of some results from polyhedral theory, a subject which provides us with the necessary combinatorial tools for the analysis of piecewise affine functions. It is way beyond the scope of this section to serve as an introduction to the beautiful and rich field of polyhedral combinatorics. Instead we have confined ourselves to the mere presentation of some notions and results which we need in the subsequent sections of this chapter. We have not included proofs of results which are well accessible in standard textbooks.

2.1.1 Convex Sets and Convex Cones

Before we introduce polyhedral sets, we present some general notions and results from convex analysis. For a set $S \subseteq \mathbb{R}^n$ we define

$$\lin S = \left\{ \sum_{i=1}^{m} \lambda_i s_i \,\middle|\, m \in \mathbb{N}, s_i \in S, \lambda_i \in \mathbb{R} \right\},$$

$$\aff S = \left\{ \sum_{i=1}^{m} \lambda_i s_i \,\middle|\, m \in \mathbb{N}, s_i \in S, \lambda_i \in \mathbb{R}, \sum_{i=1}^{m} \lambda_i = 1 \right\},$$

$$\conv S = \left\{ \sum_{i=1}^{m} \lambda_i s_i \,\middle|\, m \in \mathbb{N}, s_i \in S, \lambda_i \in \mathbb{R}_+, \sum_{i=1}^{m} \lambda_i = 1 \right\},$$

$$\cone S = \left\{ \sum_{i=1}^{m} \lambda_i s_i \,\middle|\, m \in \mathbb{N}, s_i \in S, \lambda_i \in \mathbb{R}_+ \right\}, \tag{2.1}$$

S. Scholtes, *Introduction to Piecewise Differentiable Equations*, SpringerBriefs in Optimization, DOI 10.1007/978-1-4614-4340-7_2, © Stefan Scholtes 2012

The sets $\lin S$, $\aff S$, and $\conv S$ are called the *linear, affine*, and *convex hull of S*, respectively, while the set $\cone S$ is called the *cone generated by S*. A set $S \subseteq \mathbb{R}^n$ is called *convex* if $\conv S = S$. In particular, $\lin S$, $\aff S$, and $\cone S$ are convex sets for every $S \subseteq \mathbb{R}^n$. A set $S \subseteq \mathbb{R}^n$ is called a *convex cone* if $\cone S = S$. In particular, the linear hull of a set S is a convex cone. A convex cone C is called *pointed* if it does not contain a nontrivial linear subspace, i.e., if $\lin\{x\} \subseteq C$ implies $x = 0$ for every $x \in \mathbb{R}^n$. We define the *dimension* of a convex set S to be the dimension of the affine subspace $\aff S$. A point $x \in S$ is called a *relative interior point* of S if there exists a number $\varepsilon > 0$ such that every point $y \in \aff S$ with $\|y - x\| < \varepsilon$ is contained in S. The set of all relative interior points of S, denoted by $\relint S$, is called the *relative interior* of S.

There are two important set-valued mappings which relate a closed convex set $S \subseteq \mathbb{R}^n$ to the dual space of all linear functionals on $\mathbb{R}^n$. The first mapping N_S is defined by

$$N_S(x) = \begin{cases} \{y \in \mathbb{R}^n | y^T x \geq y^T z \text{ for every } z \in S\} & \text{if } x \in S, \\ \emptyset & \text{otherwise.} \end{cases} \tag{2.2}$$

It assigns to every point $x \in S$ the set of all linear functions which achieve their maximum over S at the point x. The set $N_S(x)$ is called the *normal cone* of S at $x \in S$. We recall some of its properties:

1. The normal cone of a convex set S at $x \in S$ is a closed convex cone.
2. The normal cone is a local concept, i.e., if S and $\tilde{S}$ are convex subsets of $\mathbb{R}^n$ and U is a neighborhood of $x \in \mathbb{R}^n$ such that $S \cap U = \tilde{S} \cap U$, then $N_S(x) = N_{\tilde{S}}(x)$.
3. If the set S is a closed convex cone S, then $N_S(x) \subseteq N_S(0)$ for every $x \in S$ and the relation $N_{N_S(0)}(0) = S$ holds.

The second mapping F_S assigns to each linear functional in $\mathbb{R}^n$ the set of all maximizers $x \in S$ of the linear functional over S, i.e.,

$$F_S(y) = \{x \in S | y^T x \geq y^T z \text{ for every } z \in S\} . \tag{2.3}$$

The set $F_S(y)$ is called the *max-face* of the closed convex set S corresponding to the vector $y \in \mathbb{R}^n$. The max-faces of S have the following properties:

1. Every nonempty max-face $F_S(y)$ is convex, since it is the intersection of S with the hyperplane $\{x \in \mathbb{R}^n | y^T x = \max_{z \in S} y^T z\}$.
2. The nonempty intersection of two max-faces $F_S(y)$ and $F_S(z)$ is again a max-face of S. In fact, if $F_S(y) \cap F_S(z) \neq \emptyset$, then $F_S(y) \cap F_S(z) = F_S(y + z)$: If $\hat{x} \in F_S(y) \cap F_S(z)$ and $x \in S$ then $y^T \hat{x} \geq y^T x$ and $z^T \hat{x} \geq z^T x$ and therefore $(y + z)^T \hat{x} \geq (y + z)^T x$. Hence $\hat{x} \in F_S(y + z)$. To see the converse, suppose $\hat{x} \in F_S(y + z)$ and $\hat{x} \notin F_S(y) \cap F_S(z)$. Since $F_S(y) \cap F_S(z) \neq \emptyset$ there exists $\bar{x} \in S$ with $y^T \bar{x} \geq y^T \hat{x}$ and $z^T \bar{x} \geq z^T \hat{x}$ and as $\hat{x} \notin F_S(y) \cap F_S(z)$ one of the two inequalities holds strictly and therefore $(y + z)^T \bar{x} > (y + z)^T \hat{x}$. This contradicts $\hat{x} \in F_S(y + z)$.

3. The inclusion $x \in F_S(y)$ holds if and only $y \in N_S(x)$. This is an immediate consequence of the definition of both mappings.

A subset X of a convex set S is called an *extremal set* of S if whenever a point $x \in X$ is contained in a line segment joining two points $v, w \in S$, then both points are contained in X. If the singleton $\{x\}$ is an extremal set of a convex set S, then x is called an *extremal point* of S. If $x \neq 0$ and cone$\{x\}$ is an extremal set of a convex cone S, then cone$\{x\}$ is called an *extremal ray* of S. The vector x is called a *unit generator* of the extremal ray cone$\{x\}$ if $\|x\| = 1$.

The following results are well known from convex analysis.

Proposition 2.1.1. *1. If $S \subseteq \mathbb{R}^n$ is a compact set, then conv$S$ is compact.*
2. A convex compact set $S \subseteq \mathbb{R}^n$ is the convex hull of its extremal points.
3. A closed pointed convex cone is generated by its extremal rays.

2.1.2 Polyhedral Sets and Polyhedral Cones

The decisive property of convex sets is the *separation property* which states that for every $z \in \mathbb{R}^n$ which does not belong to the convex set $S \subseteq \mathbb{R}^n$ there exists a vector $y \in \mathbb{R}^n$ such that $y^T z \geq y^T x$ for every $x \in S$, i.e., S is contained in a halfspace which does not contain x as an interior point. In fact, every closed convex set $S \subseteq \mathbb{R}^n$ is the intersection of all halfspaces containing S, i.e., every closed convex set is the solution set of a possibly infinite system of linear inequalities. To avoid the difficulties caused by an infinite number of inequalities, one has introduced the concept of a polyhedron as the nonempty solution set of a finite system of linear inequalities. Thus a nonempty set $P \subseteq \mathbb{R}^n$ is called a *polyhedron* if there exists an $m \times n$-matrix A and an m-vector b such that $P = \{x \in \mathbb{R}^n | Ax \leq b\}$. A compact polyhedron is called a *polytope*. The solution set of a homogeneous system of linear inequalities is called a *polyhedral cone*. A polyhedral cone is thus representable as

$$C = \{x \in \mathbb{R}^n | Ax \leq 0\}, \tag{2.4}$$

where A is an $m \times n$-matrix. Clearly a polyhedron is a closed convex set, while a polyhedral cone is a closed convex cone. The following important result characterizes polyhedral cones.

Theorem 2.1.1 (Farkas–Minkowski–Weyl Theorem). *A set $C \subseteq \mathbb{R}^n$ is a polyhedral cone if and only if there exists an $l \times n$-matrix B such that*

$$C = \left\{ x \in \mathbb{R}^n | x = B^T \lambda, \lambda \in \mathbb{R}^l_+ \right\}. \tag{2.5}$$

In other words, a convex cone C is polyhedral if and only if it is generated by a finite set $\{b_1, \ldots, b_l\} \subseteq \mathbb{R}^n$. If C is pointed, then we may use the unit generators of its extremal rays as the vectors $b_1, \ldots, b_l$. The representations (2.4) and (2.5) are called the *primal* and *dual representation* of C, respectively. If $x \in C$ and C is given in dual form (2.5), then a vector $\lambda \in \mathbb{R}^m_+$ such that $x = B^T \lambda$ is called a *multiplier vector* corresponding to x.

Applying the Farkas–Minkowski–Weyl Theorem to the set cone$\{(x, 1)|x \in P\}$ $\subseteq \mathbb{R}^{n+1}$, one can show that $P \subseteq \mathbb{R}^n$ is a polytope if and only if it is the convex hull of a finite point set. In particular a polytope has only a finite number of extremal points.

Note that the pointwise addition of two nonempty sets $A, B \subseteq \mathbb{R}^n$ is defined by $A + B = \{a + b|a \in A, b \in B\}$. An important result in polyhedral theory is the decomposability of a polyhedron into the sum of a linear subspace, a pointed polyhedral cone, and a compact polyhedron.

Theorem 2.1.2 (Decomposition theorem). *If $P \subseteq \mathbb{R}^n$ is a polyhedron, then there exists a unique linear subspace $L \subseteq \mathbb{R}^n$, a unique pointed polyhedral cone $C \subseteq L^\perp$, and a compact polyhedron $Q \subseteq L^\perp$ such that $P = L + C + Q$. The set $C + Q$ necessarily coincides with $P \cap L^\perp$.*

If $P = \{x \in \mathbb{R}^n|Ax \le b\}$, then $L = \{x \in \mathbb{R}^n|Ax = 0\}$. The linear subspace L is called the *lineality space* of P, while the cone $L + C$ is called the *recession cone* of P. The recession cone of a polyhedron $P = \{x \in \mathbb{R}^n|Ax \le b\}$ is the set $L + C = \{x \in \mathbb{R}^n|Ax \le 0\}$. The polyhedron P is called *pointed* if its lineality space vanishes, or, equivalently, if its recession cone is pointed. If P is a cone, then we may choose $Q = \{0\}$ in the decomposition. Hence every polyhedral cone can be uniquely decomposed as the sum of its lineality space and a pointed polyhedral cone which is contained in the orthogonal complement of the lineality space.

The normal cones of polyhedral cones at the origin are characterized by the following useful lemma.

Lemma 2.1.1 (Farkas' Lemma). *If $C = \{x \in \mathbb{R}^n|a_i^T x \le 0, i = 1, \ldots, m\}$, then the normal cone of C at the origin is given by $N_C(0) = \text{cone}\{a_i|i = 1, \ldots, m\}$.*

In view of the Farkas–Minkowski–Weyl theorem, the latter lemma shows that the normal cone $N_C(0)$ of a polyhedral cone C is a polyhedral cone as well. It can be used to describe the normal cone of a polyhedron.

Proposition 2.1.2. *If $P = \{x \in \mathbb{R}^n|a_i^T x \le b_i, i = 1, \ldots, m\}$ and $x \in P$, then the normal cone of P at x is given by $N_P(x) = \text{cone}\{a_i|i \in \{1, \ldots, m\}, a_i^T x = b_i\}$.*

A proof of the latter proposition is easily carried out with the aid of Farkas' Lemma using the fact that the normal cone is a local concept and that the polyhedron P coincides in a neighborhood of x with the polyhedron $P(x) = \{x\} + \{y \in \mathbb{R}^n|a_i^T y \le 0, i \in \{j|a_j^T x = b_i\}\}$.

2.1.3 The Face Lattice of a Polyhedron

The collection of all max-faces of a polyhedron P together with the empty set is called the *face lattice* of P, its elements are called the *faces* of P. The term face lattice is justified by the property that the nonempty intersection of two max-faces of P is again a max-face of P and thus there is a unique minimal max-face containing two fixed max-faces of P. A nonempty face of P which does not coincide with P is called a *proper face*. The extremal points of a polyhedron P are the faces of dimension zero, while the extremal rays of a polyhedral cone are the faces of dimension 1.

To represent the nonempty faces of a polyhedron $P = \{x \in \mathbb{R}^n | Ax \le b\}$, we define for an $m \times n$-matrix A with row vectors $a_1, \ldots, a_m$ and a vector $b \in \mathbb{R}^m$ the collection of index sets

$$\mathscr{I}(A, b) = \Big\{ I \subseteq \{1, \ldots, m\} | \text{there exists a vector } x \in \mathbb{R}^n \text{ with}$$

$$a_i^T x = b_i, i \in I, a_j^T x < b_j, j \in \{1, \ldots, m\} \setminus I \Big\}. \tag{2.6}$$

and for every index set $I \subseteq \{1, \ldots, m\}$ the polyhedron

$$F_I = \Big\{ x \in \mathbb{R}^n | a_i^T x = b_i, i \in I, a_j^T x \le b_j, j \in \{1, \ldots, m\} \setminus I \Big\}. \tag{2.7}$$

Proposition 2.1.3. *If $P = \{x \in \mathbb{R}^n | Ax \le b\}$, then*

1. *A subset $F \subseteq P$ is a max-face of P if and only if there exists an index set $I \in \mathscr{I}(A, b)$ such that $F = F_I$,*
2. *Any two faces F_I, F_J corresponding to distinct index sets $I, J \in \mathscr{I}(A, b)$ are distinct,*
3. *If $I \in \mathscr{I}(A, b)$, then the relative interior of the face F_I is the set $G_I = \{x \in \mathbb{R}^n | a_i^T x = b_i, i \in I, a_j^T x < b_j, j \in \{1, \ldots, m\} \setminus I\}$,*
4. *$I \cap J \in \mathscr{I}(A, b)$ for any two index sets $I, J \in \mathscr{I}(A, b)$.*

Proof. 1. The first part is an elementary exercise. In fact, if, on the one hand, $I \in \mathscr{I}(A, b)$, then the set F_I coincides with the max-face $F_P(\sum_{i \in I} a_i)$. If, on the other hand, $F_P(y)$ is a max-face of P, then $F_P(y) = F_I$, where I is the maximal subset of $\{1, \ldots, m\}$ with the property that $a_i^T x = b_i$ for every $i \in I$ and every $x \in F_P(y)$.

2. Let $I, J \in \mathscr{I}(A, b)$ be distinct index sets. Interchanging the role of I and J if necessary, we may assume without loss of generality that there exists an index $i \in I$ with $i \notin J$. Since $J \in \mathscr{I}(A, b)$, there exists a vector $x \in F_J$ with $a_i^T x < b_i$. Thus $x \notin F_I$ which shows that $F_I \ne F_J$.

3. Recall that $x \in P$ is a relative interior point of P if there exists a positive number ε such that $y \in P$ whenever $y \in \mathrm{aff}\, P$ and $\|y - x\| < \varepsilon$. One readily verifies that

$$\mathrm{aff}\, F_I = \{x \in \mathbb{R}^n \,|\, a_i^T x = b_i, i \in I\} \tag{2.8}$$

for every $I \in \mathscr{I}(A, b)$. If, on the one hand, $x \in F_I \backslash G_I$, then $a_j^T x = b_j$ for some $j \notin I$. Since $G_I \neq \varnothing$, we may choose a vector $y \in G_I$. In particular, $a_j^T y - b_j < 0$. Defining $z(t) = x + t(y - x)$, we first conclude that $z(t) \in \mathrm{aff}\, F_I$ for every $t \in \mathbb{R}$. However, if $t < 0$, then $a_j^T z = b_j + t(a_j^T y - b_j) > b_j$ and thus $z(t) \notin P$. This shows that x is not a relative interior point of F_I. The reverse statement that every point of G_I is a relative interior point of F_I is a trivial consequence of (2.8).

4. If $v, w \in \mathbb{R}^n$ are vectors satisfying

$$a_i^T v = b_i, i \in I, a_k^T v < b_k, k \in \{1, \ldots, m\} \backslash I,$$

$$a_j^T w = b_j, j \in J, a_l^T w < b_l, l \in \{1, \ldots, m\} \backslash J,$$

then the vector $\frac{v+w}{2}$ satisfies

$$a_i^T \frac{v+w}{2} = b_i, i \in I \cap J, a_k^T \frac{v+w}{2} < b_k, k \in \{1, \ldots, m\} \backslash (I \cap J);$$

hence $I \cap J \in \mathscr{I}(A, b)$. $\square$

Since $F_I \subseteq F_J$ if and only if $I \supseteq J$, statement 1 of the latter proposition shows that the face lattice of $P = \{x \in \mathbb{R}^n \,|\, Ax \leq b\}$ and the lattice $\mathscr{I}(A, b)$ are isomorphic if the ordering relation is suitably defined.

The following result is concerned with the faces of the sum of two polyhedra.

Lemma 2.1.2. *If $Q, R \subseteq \mathbb{R}^n$ are polyhedra with $P = Q + R$, then every nonempty face of P is uniquely decomposable as the sum of a face of Q and a face of R.*

Proof. The proof is a consequence of the fact that

$$F_{P+Q}(y) = F_P(y) + F_Q(y).$$

To see this, note that

$$y^T x = \max_{p \in P} y^T p$$

$$= \max_{\substack{q \in Q \\ r \in R}} y^T q + y^T r$$

$$= \max_{q \in Q} y^T q + \max_{r \in R} y^T r,$$

and hence if $x \in P$ is represented as $x = v + w$ for some $v \in Q$, $w \in R$, then x is a maximizer of the linear function $y^T p$ over the polyhedron P if and only if v and w are maximizers of the same function over the polyhedra Q and R, respectively. This shows that the face of P corresponding to the maximizers of $y^T p$ is the sum of the faces of Q and R corresponding to the maximizers of the same linear function. $\square$

As an immediate consequence of the decomposition theorem and the latter lemma, we obtain the following corollary.

Corollary 2.1.1. *The lineality space of a face of a polyhedron coincides with the lineality space of the polyhedron.*

2.1.4 Comments and References

For comprehensive treatments of polyhedral theory we refer the interested reader to Grünbaum's monograph [22] and to the recent book of Ziegler [82]. A comprised account with an emphasis on linear programming applications can be found in Schrijver's text [73]. The standard reference for the more general results from convex analysis is Rockafellar's monograph [67].

The statements of Proposition 2.1.1 are stated as Theorem 17.2 and Theorem 18.5 in [67], while the Farkas–Minkowski–Weyl Theorem, the Decomposition Theorem, and Farkas' Lemma are proved as Corollaries 7.1a, 7.1b, and 7.1d, respectively, in [73].

2.2 Basic Notions and Properties

We proceed by setting the basic notions for the analysis of piecewise affine function. We start with a precise definition. A continuous function $f : \mathbb{R}^n \to \mathbb{R}^m$ is called *piecewise affine* if there exists a finite set of affine functions $f_i(x) = A^i x + b^i$, $i = 1, \ldots, k$, such that the inclusion $f(x) \in \{f_1(x), \ldots, f_k(x)\}$ holds for every $x \in \mathbb{R}^n$. The affine functions $f_i(x) = A^i x + b^i$, $i = 1, \ldots, k$, are called *selection functions*, the set of pairs (A^i, b^i), $i = 1, \ldots, k$, is called a collection of *matrix-vector pairs* corresponding to f. The function f is called *piecewise linear* if there exists a corresponding set of linear selection functions.

Similar as in the affine case, the superposition $(f \circ g)(x) = f(g(x))$ of two piecewise affine functions $f : \mathbb{R}^n \to \mathbb{R}^m$ and $g : \mathbb{R}^k \to \mathbb{R}^n$ is again piecewise affine. In fact, the set of all possible superpositions of selection functions corresponding to f and g, respectively, forms a collection of selection functions for $f \circ g$. It is easily checked that a continuous function $f : \mathbb{R}^n \to \mathbb{R}^m$ is piecewise affine if and only if all real-valued coordinate functions $f_1, \ldots, f_m : \mathbb{R}^n \to \mathbb{R}$ are piecewise affine. The following result provides a useful characterization of piecewise linear functions.

Proposition 2.2.1. *A piecewise affine function* f *is piecewise linear if and only if it is positively homogeneous, i.e.,* $f(\alpha x) = \alpha f(x)$*, for every nonnegative real number* α*.*

Proof. Let $f: \mathbb{R}^n \to \mathbb{R}^m$ be a positively homogeneous piecewise affine function with matrix-vector pairs (A^i, b^i), $i = 1, \ldots, k$. If $k = 1$, then the claim is trivial. So suppose that $k \geq 2$ and that $b^k \neq 0$. We will show that for every $x \in \mathbb{R}^n$ with $f(x) = A^k x + b^k$ there exists an index $i \neq k$ such that $f(x) = A^i x + b^i$. In fact, since f is positively homogeneous, the identity $f(x) = A^k x + b^k$ implies that

$$f(\alpha x) = \alpha f(x) = \alpha A^k x + \alpha b^k \neq A^k \alpha x + b^k$$

for every positive number $\alpha \neq 1$. Hence $f(\alpha x) = A^k \alpha x + b^k$ if and only if $\alpha = 1$ and thus the continuity of f shows that there exists another matrix–vector pair (A^i, b^i) such that $f(x) = A^i x + b^i$. Hence the matrix-vector pairs (A^i, b^i), $i = 1, \ldots, k-1$, form a collection of matrix–vector pair corresponding to f and thus an induction argument completes the proof that a positively homogeneous piecewise affine function is piecewise linear. □

It is a matter of plane geometry to check that a real-valued piecewise linear function $\phi: \mathbb{R} \to \mathbb{R}$ of a single variable is positively homogeneous. To prove the general case, suppose $f: \mathbb{R}^n \to \mathbb{R}^m$ is piecewise linear with corresponding matrices $A^1, \ldots, A^k$. Fix a vector $x \in \mathbb{R}^n$ and consider the function $\phi(t) = f(tx)$. If A^i_j denotes the jth row of the matrix A^i, then the component functions $\phi_j: \mathbb{R} \to \mathbb{R}$ are piecewise linear with selection functions $A^1_j tx, \ldots, A^k_j tx$. Hence the component functions $\phi_j: \mathbb{R} \to \mathbb{R}$ are piecewise linear and thus the function ϕ is positively homogeneous, which shows that $f(\alpha x) = \phi(\alpha) = \alpha \phi(1) = \alpha f(x)$ for every $\alpha \geq 0$. □

2.2.1 Representations of Piecewise Affine Functions

Typical examples of real-valued piecewise affine functions are the pointwise maxima or minima of a finite set of affine functions, or, more generally, functions which are built up by superpositions of finitely many maximum or minimum functions. It is rather surprising that every real-valued piecewise affine function is in fact a function of this type.

Proposition 2.2.2. *If* $f: \mathbb{R}^n \to \mathbb{R}$ *is a piecewise affine with affine selection functions* $f_1(x) = a_1^T x + b_1, \ldots, f_k(x) = a_k^T x + b_k$*, then there exists a finite number of index sets* $M_1, \ldots, M_l \subseteq \{1, \ldots, k\}$ *such that*

$$f(x) = \max_{1 \leq i \leq l} \min_{j \in M_i} a_i^T x + b_i.$$

Proof. In the first part of the proof we construct a max–min function of the required form which is then shown to coincide with f. Removing redundant selection functions, if necessary, we may assume without loss of generality that the selection functions are mutually distinct, i.e.,

$$(a_i, b_i) \neq (a_j, b_j) \tag{2.9}$$

for every $i \neq j$. For a permutations π of the numbers $\{1, \ldots, k\}$ we define the set

$$M(\pi) = \left\{ x \in \mathbb{R}^n \, | \, a_{\pi(1)}^T x + b_{\pi(1)} \leq \cdots \leq a_{\pi(k)}^T x + b_{\pi(k)} \right\}.$$

Note that $M(\pi)$ is the set of solutions of the inequalities

$$(a_{\pi(1)} - a_{\pi(2)})^T x \leq -b_{\pi(1)} + b_{\pi(2)}$$
$$(a_{\pi(2)} - a_{\pi(3)})^T x \leq -b_{\pi(2)} + b_{\pi(3)}$$
$$\vdots$$
$$(a_{\pi(k-1)} - a_{\pi(k)})^T x \leq -b_{\pi(k-1)} + b_{\pi(k)}$$

and thus either empty or a convex polyhedron. Let Π be the set of all permutations π of the numbers $\{1, \ldots, k\}$ with the property $\mathrm{int} M(\pi) \neq \emptyset$. Note that

$$\bigcup_{\pi \in \Pi} M(\pi) = \mathbb{R}^n, \tag{2.10}$$

since by definition the union of all polyhedra $M(\pi)$, where π is a permutation of the numbers $\{1, \ldots, k\}$, covers $\mathbb{R}^n$ and the removal of polyhedra with empty interior does not affect this covering property. In view of (2.9) and part 3 of Proposition 2.1.3, the interior of the polyhedron $M(\pi)$ is given by the solution set of the strict inequalities, i.e.,

$$\mathrm{int} M(\pi) = \left\{ x \in \mathbb{R}^n \, | \, a_{\pi(1)}^T x + b_{\pi(1)} < \cdots < a_{\pi(k)}^T x + b_{\pi(k)} \right\}. \tag{2.11}$$

Since f is continuous and the selection functions are mutually distinct, we can thus find for every $\pi \in \Pi$ a unique index $i_\pi \in \{1, \ldots, k\}$ such that

$$f(x) = a_{\pi(i_\pi)}^T x + b_{\pi(i_\pi)}$$

for every $x \in M(\pi)$. With the aid of the indices i_π, we define the max–min function

$$g(x) = \max_{\pi \in \Pi} \min_{i \in \{i_\pi, \ldots, k\}} a_{\pi(i)}^T x + b_{\pi(i)}, \tag{2.12}$$

which is a function of the form required by the assertion. In order to prove that the functions f and g coincide, we choose an arbitrary vector $x_0 \in \mathbb{R}^n$. Equation (2.10) shows that there exists a permutation $\hat{\pi} \in \Pi$ with $x_0 \in M(\hat{\pi})$. The definitions of the set $M(\hat{\pi})$ and the index $i_{\hat{\pi}}$ yield

$$f(x_0) = a^T_{\hat{\pi}(i_{\hat{\pi}})} x_0 + b_{\hat{\pi}(i_{\hat{\pi}})} = \min_{i \in \{i_{\hat{\pi}}, \dots, k\}} a^T_{\hat{\pi}(i)} x_0 + b_{\hat{\pi}(i)}$$

and hence

$$f(x_0) \leq \max_{\pi \in \Pi} \min_{i \in \{i_\pi, \dots, k\}} a^T_{\pi(i)} x_0 + b_{\pi(i)} = g(x_0). \tag{2.13}$$

$\square$

In order to prove the reverse inequality, we show that for every $\pi \in \Pi$ the inequality

$$f(x_0) \geq \min_{i \in \{i_\pi, \dots, k\}} a^T_{\pi(i)} x_0 + b_{\pi(i)} \tag{2.14}$$

holds. Suppose the contrary, i.e.,

$$f(x_0) < \min_{i \in \{i_\pi, \dots, k\}} a^T_{\pi(i)} x_0 + b_{\pi(i)}$$

for some $\pi \in \Pi$. To simplify the exposition, we renumber the selection functions in such a way that π is the identity, i.e., $\pi(i) = i$ for $i = 1, \dots, k$ and hence

$$f(x_0) < \min_{i \in \{i_\pi, \dots, k\}} a^T_i x_0 + b_i. \tag{2.15}$$

Since $\pi \in \Pi$, we can find a vector

$$y_0 \in \mathrm{int} M(\pi) = \left\{ x \in \mathbb{R}^n \mid a^T_1 x + b_1 < \cdots < a^T_k x + b_k \right\}. \tag{2.16}$$

Since f is piecewise affine, the image of the line segment $[x_0, y_0]$ is a polygonal path. In fact, there exists a finite number of indices i_j, $j = 0, \dots, l$, and corresponding vectors $x_j \in [x_0, y_0]$ such that

$$f(x) = a^T_{i_j} x + b_{i_j} \text{ for every } j \in \{0, \dots, l\} \text{ and every } x \in [x_j, x_{j+1}], \tag{2.17}$$

where $x_{l+1} = y_0$. Next, we prove that for every $j \in \{0, \dots, l\}$ the inequalities

$$a^T_{i_j} x_j + b_{i_j} < a^T_i x_j + b_i, i \in \{i_\pi, \dots, k\},$$

$$a^T_{i_j} y_0 + b_{i_j} < a^T_i y_0 + b_i, i \in \{i_\pi, \dots, k\} \tag{2.18}$$

hold. This is done by induction over j. For $j = 0$, the first set of inequalities is an immediate consequence of (2.15). In fact, (2.15) implies that $i_0 \notin \{i_\pi, \dots, k\}$, which yields the second set of inequalities in view of (2.16). Now we assume that the inequalities (2.18) hold for the index j, and prove their validity for the index $j + 1$. One readily deduces from the validity of the inequalities for the index j that

$$a_{i_j}^T x + b_{i_j} < a_i^T x + b_i \tag{2.19}$$

for every $x \in [x_j, y_0]$ and every $i \in \{i_\pi, \dots, k\}$. Since in view of (2.17) the identity

$$a_{i_{j+1}}^T x_{j+1} + b_{i_{j+1}} = a_{i_j}^T x_{j+1} + b_{i_j}, \tag{2.20}$$

holds, the inequalities (2.19) applied to $x = x_{j+1}$ yield the first set of inequalities in (2.18) for the index $j + 1$. Furthermore, (2.19) and (2.20) show that $i_{j+1} \notin \{i_\pi, \dots, k\}$ and thus (2.16) yields the validity of the second set of inequalities. Setting $j = l$ and recalling that $x_{j+1} = y_0$, we may apply (2.17) and the fact that $y_0 \in M(\pi)$ to obtain

$$f(y_0) = a_{i_l}^T y_0 + b_{i_l} < a_{i_\pi}^T y_0 + b_{i_\pi} = f(y_0),$$

which is a contradiction. Hence (2.15) does not hold and thus (2.14) holds for every $\pi \in \Pi$, which completes the proof of the proposition. $\square$

The latter representation result shows that the component functions of every piecewise affine function $f : \mathbb{R}^n \to \mathbb{R}^m$ can be represented as max–min functions. Such a representation is called a *max–min representation* of f.

While the max–min representation has many theoretical benefits, a severe drawback of the max–min form constructed in the latter proof is the large number of minimum functions appearing in it. Of course the max–min representation of a piecewise affine function is generally not unique. However, the construction of a handy max–min form is often very difficult, if at all possible. The reader may for instance try to construct a tractable max–min form for the Euclidean projection onto a polyhedron (cf. Sect. 2.4). In practice, a piecewise affine function is often specified by providing a finite collection Σ of subsets of $\mathbb{R}^n$ on each of which the function coincides with an affine function together with the corresponding selection functions. Suppose a set of selection functions $A^i x + b^i$, $i = 1, \dots, k$, is given, and consider the sets $\sigma_i = \{x \in \mathbb{R}^n \,|\, f(x) = A^i x + b^i\}$, $i = 1, \dots, k$. Since f is continuous, the sets σ_i are closed and since f is piecewise affine, the union of all these sets covers $\mathbb{R}^n$. This covering property is not affected if we remove from the collection all sets σ_i with empty interior. Moreover, if the interiors of two sets σ_i and σ_j have a point in common, then the affine functions $A^i x + b^i$ and $A^j x + b^j$ coincide. Hence, if we require the selection functions to be mutually distinct, the collection Σ of all sets σ_i with nonempty interior is a so-called *partition* of $\mathbb{R}^n$, i.e., every $\sigma \in \Sigma$ is a closed subset of $\mathbb{R}^n$ with nonempty interior, no two distinct sets $\sigma, \tilde{\sigma} \in \Sigma$ have a common interior point, and the union of all sets $\sigma \in \Sigma$ covers $\mathbb{R}^n$. We say that a partition Σ corresponds to the piecewise affine function f if f coincides with an affine function on every set $\sigma \in \Sigma$.

The collection Σ just constructed is thus a partition of $\mathbb{R}^n$ corresponding to f. Note that the resulting collection of selection functions is the smallest possible and that it is uniquely determined. We will call this collection and the corresponding partition the *minimal collection* of selection functions and the *minimal partition*, respectively, corresponding to f. Although minimal in number, the sets in the minimal partition lack some desirable properties, like convexity or even connectedness. For some purposes, however, such additional structural properties of the partition are very useful. If $P \subseteq \mathbb{R}^n$ is a convex polyhedron and Σ is a finite collection of convex polyhedra in $\mathbb{R}^n$, then Σ is called a *polyhedral subdivision* of P if

1. All polyhedra in Σ are subsets of P
2. The dimension of the polyhedra in Σ coincides with the dimension of P
3. The union of all polyhedra in Σ covers P
4. The intersection of any two polyhedra in Σ is either empty or a common proper face of both polyhedra

The polyhedral subdivision Σ of P is called *pointed* if the lineality space of every polyhedron vanishes. If P is a polyhedral cone and all polyhedra in Σ are polyhedral cones, then Σ is called a *conical subdivision* of P. The collection of all j-dimensional faces of the polyhedra $\sigma \in \Sigma$, denoted by Σ_j, is called the j-*skeleton* of Σ. For a collection Σ of subsets of $\mathbb{R}^n$, we denote by $|\Sigma|$ the union of all sets in Σ. The set $|\Sigma|$ is called the *carrier* of Σ.

It is not clear a priori, whether every piecewise affine function can be endowed with a corresponding polyhedral subdivision of $\mathbb{R}^n$. The following proposition shows that this is indeed the case.

Proposition 2.2.3. *Every piecewise affine (piecewise linear) function $f : \mathbb{R}^n \to \mathbb{R}^m$ admits a corresponding polyhedral (conical) subdivision of $\mathbb{R}^n$.*

Proof. Let $f = (f_1, \ldots, f_m)$, where the functions f_j are the real-valued component functions of f. Suppose the affine selection functions of the component function f_j are the functions $(v_i^j)^T x + \alpha_i^j$, $i = 1, \ldots, k_j$, where we assume that $(v_i^j, \alpha_i^j) \neq (v_{i'}^j, \alpha_{i'}^j)$ for every $i \neq i'$. For a vector $\pi = (\pi_1, \ldots, \pi_m)$ of permutations π_i of the numbers $\{1, \ldots, k_j\}$ we define the set

$$M(\pi) = \left\{ x \in \mathbb{R}^n | (v_{\pi_j(1)}^j)^T x + \alpha_{\pi_j(1)}^j \leq \cdots \leq (v_{\pi_j(k_j)}^j)^T x + \alpha_{\pi_j(k_j)}^j, 1 \leq j \leq m \right\}.$$

We claim that the collection of all sets $M(\pi)$ which have nonempty interior forms a polyhedral subdivision of $\mathbb{R}^n$ corresponding to the piecewise affine function f. Clearly every set $M(\pi)$ is a polyhedron since it can be easily transformed into the solution set of a finite system of linear inequalities. Since the selection functions of the components are mutually distinct, the interior of a polyhedron $M(\pi)$ is given by

$$\text{int} M(\pi) = \left\{ x \in \mathbb{R}^n | (v_{\pi_j(1)}^j)^T x + \alpha_{\pi_j(1)}^j < \cdots < (v_{\pi_j(k_j)}^j)^T x + \alpha_{\pi_j(k_j)}^j, 1 \leq j \leq m \right\}.$$

Hence every component function coincides on $\mathrm{int}M(\pi)$ with a single selection function and thus f coincides with an affine function on $M(\pi)$ if the latter set has a nonempty interior. It remains to prove that the collection of all polyhedra with nonempty interior constitutes a polyhedral subdivision of $\mathbb{R}^n$. Note first that the union of all sets $M(\pi)$ covers $\mathbb{R}^n$ and that this covering property is not affected if the polyhedra with empty interior are removed. It thus remains to prove that any two distinct sets $M(\pi)$ and $M(\tilde{\pi})$ with nonempty intersection share a common proper face. It follows immediately from the definition of $M(\pi)$ that the latter set can be represented by

$$M(\pi) = \left\{ x \in \mathbb{R}^n \,|\, s_{i,i',j}(\pi)(v_i^j - v_{i'}^j)^T x \le s_{i,i',j}(\pi)(-\alpha_i^j + \alpha_{i'}^j), \right.$$

$$\left. 1 \le j \le m, 1 \le i, i' \le k_j, i \ne i' \right\}.$$

where

$$s_{i,i',j}(\pi) = \begin{cases} 1 & \text{if } \pi_j^{-1}(i) < \pi_j^{-1}(i'), \\ -1 & \text{if } \pi_j^{-1}(i) > \pi_j^{-1}(i'). \end{cases}$$

Note that an inequality representation of $M(\pi) \cap M(\tilde{\pi})$ is obtained from the representation of $M(\pi)$ by turning inequalities into equalities whenever the relation $s_{i,i',j}(\pi) = -s_{i,i',j}(\tilde{\pi})$ holds. Hence $M(\pi) \cap M(\tilde{\pi})$ is a face of $M(\pi)$. Moreover, since for any two distinct permutation vectors π and $\tilde{\pi}$ there exists at least one index triple (i, i', j) with $s_{i,i',j}(\pi) = -s_{i,i',j}(\tilde{\pi})$, and since the interior of $M(\pi)$ is the solution set of the strict inequality system, we conclude that the face is proper, provided that $\mathrm{int}M(\pi) \ne \emptyset$. $\square$

If f is piecewise linear, then the we may choose $\alpha_i^j = 0$ for every $j = 1, \ldots, m$ and every $i = 1, \ldots, k_j$ and thus the sets $M(\pi)$ are polyhedral cones. $\square$

The next property of polyhedral subdivisions is a consequence of the decomposition theorem for polyhedra.

Proposition 2.2.4. *If Σ is a polyhedral subdivision of a polyhedron $P \subseteq \mathbb{R}^n$, then all polyhedra $\sigma \in \Sigma$ have the same lineality space.*

Proof. Consider first two polyhedra $\sigma, \tilde{\sigma} \in \Sigma$ with nonempty intersection and suppose $\sigma = L + Q$, $\tilde{\sigma} = \tilde{L} + \tilde{Q}$, where L and $\tilde{L}$ are the lineality spaces of σ and $\tilde{\sigma}$, respectively, and Q and $\tilde{Q}$ are pointed polyhedra. Since the partition is a polyhedral subdivision, σ and $\tilde{\sigma}$ intersect in a common face $F = L + G = \tilde{L} + \tilde{G}$, where G and $\tilde{G}$ are faces of Q and $\tilde{Q}$, respectively. As stated in Corollary 2.1.1, the lineality space of a polyhedron coincides with the lineality space of its faces; hence G and $\tilde{G}$ are pointed polyhedra. The uniqueness of the lineality space of a polyhedron thus shows that L and $\tilde{L}$ coincide, i.e., if two polyhedra intersect in a common proper face, then their lineality spaces coincide. Suppose the polyhedral subdivision Σ is divided into two subsets, the first one containing polyhedra with lineality space L and the other one containing polyhedra with lineality spaces different from L. Then the above argument shows that any two polyhedra from different subsets have empty intersection. Since the union of both subsets is closed, the polyhedron P is the union of two closed sets with nonempty intersection. However, this contradicts the convexity of P unless one of the sets is empty. Hence the lineality spaces of all polyhedra in the subdivision coincide. $\square$

For short, we call the common lineality space of the polyhedra of a polyhedral subdivision Σ the *lineality space* of Σ.

It is sometimes helpful to work with pointed subdivisions. The final result of this section shows that every subdivision can be subdivided into a pointed subdivision.

Proposition 2.2.5. *If Σ is a polyhedral (conical) subdivision of polyhedron (polyhedral cone) $P \subseteq \mathbb{R}^n$, then there exists a pointed polyhedral (conical) subdivision $\tilde{\Sigma}$ of P such that every polyhedron $\tilde{\sigma} \in \tilde{\Sigma}$ is contained in some polyhedron $\sigma \in \Sigma$.*

Proof. Due to the decomposition theorem and Proposition 2.2.4, every polyhedron $\sigma \in \Sigma$ can be decomposed as $\sigma = L + \hat{\sigma}$, where L is the lineality space of σ and $\hat{\sigma} = \sigma \cap L^\perp$. Since any two polyhedra $\sigma_i, \sigma_j \in \Sigma$ intersect in a common face, so do the polyhedra $\hat{\sigma}_i$ and $\hat{\sigma}_j$. Next partition the subspace L into pointed cones such that any two cones intersect in a common proper face. Such a partition certainly exists since the collection of orthants in $\mathbb{R}^{\dim L}$ can be mapped onto L by means of a linear function mapping $\mathbb{R}^{\dim L}$ one-to-one onto L. Let $c_1, \ldots, c_k$ be the cones which partition the linear subspace L and let $\tilde{\Sigma}$ be the collection of all polyhedra of the form $\sigma_i \cap L^\perp + c_j$, where $\sigma_i \in \Sigma$ and $j \in \{1, \ldots, k\}$. Clearly the lineality spaces of the polyhedra in $\tilde{\Sigma}$ vanish. Using the fact that $\hat{\sigma}_i = \sigma_i \cap L^\perp$ and c_j are contained in mutually orthogonal subspaces, it is not difficult to check that two polyhedra $\hat{\sigma}_i + c_j$ and $\hat{\sigma}_p + c_q$ have nonempty intersection if and only if the polyhedra $\hat{\sigma}_i$ and $\hat{\sigma}_p$ have nonempty intersection and that in this case

$$(\hat{\sigma}_i + c_j) \cap (\hat{\sigma}_p + c_q) = \hat{\sigma}_i \cap \hat{\sigma}_p + c_j \cap c_q.$$

Moreover, since $\hat{\sigma}_i \cap \hat{\sigma}_p$ and $c_j \cap c_q$ are faces of $\hat{\sigma}_i$ and c_j, there exist two vectors $l \in L$ and $l' \in L^\perp$ with $F_{\hat{\sigma}_i}(l') = \hat{\sigma}_i \cap \hat{\sigma}_p$ and $F_{c_j}(l) = c_j \cap c_q$. Using the mutual orthogonality of L and $L^\perp$, the definition of the max-faces yields

$$F_{(\hat{\sigma}_i + c_j)}(l + l') = \hat{\sigma}_i \cap \hat{\sigma}_p + c_j \cap c_q.$$

Hence $\tilde{\Sigma}$ is a pointed polyhedral subdivision of P. It should also be clear that the collection $\tilde{\Sigma}$ is a conical subdivision if Σ is conical. $\square$

2.2.2 Local Approximations of Piecewise Affine Functions

Many problems that we encounter in the sequel are local in nature, i.e., they are concerned only with the behavior of a function in a neighborhood of a given point. For this purpose it is convenient to have a local approximation concept at hand.

Suppose $f: \mathbb{R}^n \to \mathbb{R}^m$ is a piecewise affine function with corresponding polyhedral subdivision Σ and let $x_0 \in \mathbb{R}^n$ be some fixed point. Define

$$\Sigma(x_0) = \{\sigma \in \Sigma \mid x_0 \in \sigma\}, \tag{2.21}$$

Obviously $x_0 \in \mathrm{int}|\Sigma(x_0)|$ and hence there exists an open convex neighborhood U of zero such $\{x_0\} + U \subseteq |\Sigma(x_0)|$. In particular, we can find for every vector $v \in \mathbb{R}^n$ a real number $\alpha_0 > 0$ such that $\alpha_0 v \in U$. Hence there exists a polyhedron $\sigma \in \Sigma(x_0)$ such that $x_0 + \alpha_0 v \in \sigma$ and thus the convexity of σ implies that $x_0 + \alpha v \in \sigma$ for every $\alpha \in [0, \alpha_0]$. Since f coincides on σ with an affine function, say $f(x) = Ax + b$ for every $x \in \sigma$, we thus obtain

$$\lim_{\substack{\alpha \to 0 \\ \alpha > 0}} \frac{f(x_0 + \alpha v) - f(x_0)}{\alpha} = Av.$$

Hence, the function

$$f'(x_0; v) = \lim_{\substack{\alpha \to 0 \\ \alpha > 0}} \frac{f(x_0 + \alpha v) - f(x_0)}{\alpha}$$

is well defined and positively homogeneous. In fact, it coincides on U with the function

$$\hat{f}(v) = f(x_0 + v) - f(x_0) \tag{2.22}$$

which is continuous. Since a positively homogeneous function is continuous whenever it is continuous in a neighborhood of the origin, the function $f'(x_0; .)$ is continuous. We have seen above that $f'(x_0; v)$ achieves only values $A^i v$, where A^i is a matrix corresponding to an affine function $A^i x + b^i$ which coincides with f on some polyhedron $\sigma \in \Sigma$ with $x_0 \in \sigma$; whence $f'(x_0; .)$ is piecewise linear. The function $f'(x_0; .)$ is called the *B-derivative* of f at x_0. In view of (2.22), the B-derivative contains all local information about the piecewise affine function f since the values $f(x_0) + f'(x_0; x - x_0)$ and $f(x)$ coincide in a neighborhood of x_0.

A conical subdivision of $\mathbb{R}^n$ corresponding to $f'(x_0; .)$ is given by the collection

$$\Sigma'(x_0) = \{\mathrm{cone}(\sigma - \{x_0\}) \mid \sigma \in \Sigma(x_0)\}. \tag{2.23}$$

The latter collection of polyhedral cones is called the *localization* of Σ at x_0. The subdivision property is easily verified using the fact that a subset $F' \subseteq \mathbb{R}^n$ is a face of $\mathrm{cone}(\sigma - \{x_0\})$ if and only if $F' = \mathrm{cone}(F - \{x_0\})$, where F is a face of σ with $x_0 \in F$. Moreover, one easily checks that the dimension of the lineality space of $\Sigma'(x_0)$ is p if x_0 is contained in the relative interior of a p-dimensional face of some polyhedron $\sigma \in \Sigma$.

We comprise the above observations in the following proposition.

Proposition 2.2.6. *Let $f: \mathbb{R}^n \to \mathbb{R}^m$ be a piecewise affine function, Σ be a corresponding polyhedral subdivision of $\mathbb{R}^n$, and let $x_0 \in \mathbb{R}^n$.*

1. *The collection $\Sigma'(x_0)$ is a conical subdivision of $\mathbb{R}^n$. If $x_0 \in \Sigma_p \setminus \Sigma_{p-1}$, then the dimension of the lineality space of $\Sigma'(x_0)$ is p.*
2. *The B-derivative $f'(x_0; .)$ is a piecewise linear function with corresponding conical subdivision $\Sigma'(x_0)$.*
3. *If f coincides with the affine function $Ax + b$ on the polyhedron $\sigma \in \Sigma(x_0)$, then $f'(x_0; v) = Av$ for every $v \in \text{cone}(\sigma - \{x_0\})$.*
4. *The identity $f(x) = f(x_0) + f'(x_0; x - x_0)$ holds for every $x \in |\Sigma(x_0)|$.*

2.2.3 Lipschitz Continuity of Piecewise Affine Functions

An important property of piecewise affine functions is their Lipschitz continuity. Recall that a function $f : \mathbb{R}^n \to \mathbb{R}^m$ is called Lipschitz continuous, if there exists a constant L such that

$$\|f(x) - f(y)\| \leq L\|x - y\|,$$

for every $x, y \in \mathbb{R}^n$. A suitable constant L is called a Lipschitz constant of the function f. In the case of a linear function, the smallest Lipschitz constant serves as a norm, the so-called operator norm, i.e., if A is an $m \times n$-matrix, then

$$\|\!|A|\!\| = \max_{x \neq y} \frac{\|Ax - Ay\|}{\|x - y\|} = \max_{z \neq 0} \frac{\|Az\|}{\|z\|}.$$

The number $\|\!|A|\!\|$ is also a Lipschitz constant for the affine function $Ax + b$. The piecewise affine functions inherit the Lipschitz property from the affine functions.

Proposition 2.2.7. *Every piecewise affine function $f : \mathbb{R}^n \to \mathbb{R}^m$ is Lipschitz continuous. If $(A^1, b^1), \ldots, (A^k, b^k)$, is a collection of matrix-vector pairs corresponding to f, then $\max\{\|\!|A^1|\!\|, \ldots, \|\!|A^k|\!\|\}$ is a Lipschitz constant for f.*

Proof. Let Σ be a polyhedral subdivision of $\mathbb{R}^n$ corresponding to f, let $x, y \in \mathbb{R}^n$, and let $[x, y]$ be the line segment joining x and y. Since the nonempty intersection of a line segment with a polyhedron is either a singleton or again a line segment, there exist numbers $0 = \alpha_0 \leq \alpha_1 \leq \cdots \leq \alpha_m = 1$ such that each line segment $[x + \alpha_i(x - y), x + \alpha_{i+1}(x - y)]$, $i = 0, \ldots, m - 1$, is contained in some fixed polyhedron σ_i. Since f coincides on σ_i with an affine function, say $f(x) = A^i x + b^i$ for every $x \in \sigma_i$, we conclude

$$\|f(x) - f(y)\| \leq \sum_{i=0}^{m-1} \|f(x + \alpha_i(x - y)) - f(x + \alpha_{i+1}(x - y))\|$$

$$= \sum_{i=0}^{m-1} (\alpha_{i+1} - \alpha_i)\|A^i(x - y)\|$$

$$\leq \sum_{i=0}^{m-1} (\alpha_{i+1} - \alpha_i) \|A^i\| (\|x - y\|)$$

$$\leq \left(\max_{1 \leq i \leq m-1} \|A^i\| \right) \|x - y\|. \qquad \square$$

Similarly as in the linear case, the smallest Lipschitz constant of a piecewise
linear function may serve as a norm on the linear space of piecewise linear functions.

2.2.4 Comments and References

Piecewise affine functions are usually studied within the framework of piecewise
linear topology as simplicial or cellular maps (cf. e.g. [26, 68]). In fact, in the
literature piecewise affine mappings are only defined with respect to a corresponding
polyhedral subdivision (cf., e.g., [16,20,30,33,59,70]). However, for our purposes,
the definition presented here is best suited. In particular, there is no problem to define
the superposition of two piecewise affine mappings. In view of Proposition 2.2.3
(cf. [72]) our definition of piecewise affine functions is equivalent to the definitions
given in the literature.

A different proof of the max–min representation result of Proposition 2.2.2 can be
found in [3]. The latter article contains also a uniqueness condition for the max–min
form given in Proposition 2.2.2. The results of Propositions 2.2.4 and 2.2.6 can
be found in the article [16] of Eaves and Rothblum. The Lipschitz continuity of
piecewise affine functions (cf. Proposition 2.2.7) is proved in the paper [20] of
Fujisawa and Kuh.

Some authors have extended the definition of piecewise affine functions by
replacing the finiteness of the number of selection functions by a local finiteness
condition (cf. e.g., [33]). We do not treat this case since the piecewise affine
functions appearing in applications are almost always built up by a finite number
of selection functions.

2.3 Piecewise Affine Homeomorphisms

After the introduction of the basic notions, we proceed to the main topic of this
chapter which is the study of the homeomorphism problem for piecewise affine
functions. We begin with some definitions. If $X \subseteq \mathbb{R}^n$, $Y \subseteq \mathbb{R}^m$, and $f : X \to Y$
is a continuous function on X, then f is called a *homeomorphism* if for every
$y \in Y$ there exists a unique solution $x = f^{-1}(y)$ to the equation $f(x) = y$
and the *inverse function* $f^{-1} : Y \to X$ is continuous. In this case, we say that f
maps X homeomorphically onto Y. The function $f : X \to Y$ is called a *local*

homeomorphism at a point $x \in X$ if there exist neighborhoods U of x and V of $f(x)$ such that f maps $U \cap X$ homeomorphically onto $V \cap Y$. If f is a local homeomorphism at every point $x \in X$, then f is called a *local homeomorphism*.

In the sequel we will develop conditions which ensure that a piecewise affine function $f : \mathbb{R}^n \to \mathbb{R}^n$ is a homeomorphism. By definition, a homeomorphism $f : \mathbb{R}^n \to \mathbb{R}^n$ is characterized by the following three features:

Injectivity: Every image vector $y \in \mathbb{R}^n$ has at most one preimage $x \in \mathbb{R}^n$.
Surjectivity: Every image vector $y \in \mathbb{R}^n$ has at least one preimage $x \in \mathbb{R}^n$.
Openness: The image of any open subset of $\mathbb{R}^n$ is an open subset of $\mathbb{R}^n$.

Our restriction to functions mapping $\mathbb{R}^n$ into a space of the same dimension is necessary in view of Brouwer's theorem on the invariance of dimension which implies that there is no homeomorphism mapping $\mathbb{R}^n$ into $\mathbb{R}^m$ unless $n = m$. However, in the case of a piecewise affine function it is not even necessary to appeal to the latter theorem. In fact, if $f : \mathbb{R}^n \to \mathbb{R}^m$ is piecewise affine and $A^i x + b^i$ is an affine selection function from the minimal collection of selection functions corresponding to f, then the set $\sigma_i = \{x \in \mathbb{R}^n | f(x) = A^i x + b^i\}$ has nonempty interior. If f is a homeomorphism, then the interior of σ_i is mapped homeomorphically onto $f(\text{int}\sigma_i)$ which implies that the affine function $A^i x + b^i$ maps an open set homeomorphically onto an open set. Elementary linear algebra thus shows that $n = m$.

An important property of affine homeomorphisms is the fact that their inverse functions are affine as well. This can readily be generalized to piecewise affine functions.

Proposition 2.3.1. *A piecewise affine homeomorphism has a piecewise affine inverse function.*

Proof. We have already argued above that every affine function $A^i x + b^i$ in the minimal collection of selection functions corresponding to a piecewise affine homeomorphism f is itself a homeomorphism, i.e., the matrix A^i is nonsingular. So if $f(x) = A^i x + b^i = y$, then $x = (A^i)^{-1} y + (A^i)^{-1}(-b^i)$. This shows that for every y in the image space there exists an affine function of the above type such that $f^{-1}(y)$ coincides with the value of this affine function at y. The result thus follows from the fact that the inverse function of a homeomorphism is by definition a continuous functions.   □

In this section we will be mainly concerned with the problem, how to decide whether a given piecewise affine function is a homeomorphism or not. We try to keep our exposition as close as possible to the well-known special case of affine function. In fact, most of the following results are generalizations of properties of affine homeomorphism.

2.3.1 Coherently Oriented Piecewise Affine Functions

A particularly nice property of an affine function $f : \mathbb{R}^n \to \mathbb{R}^n$ is the fact that any one of the characteristic features surjectivity, injectivity, or openness implies that f is a homeomorphism. Unfortunately, this property is not inherited by the class of piecewise affine functions. In fact, we will show in this section that in the piecewise affine case injectivity implies openness which again implies surjectivity, but that none of the reverse implications hold. Hence injectivity is the key property of a piecewise affine homeomorphism. The fact that injectivity implies openness may be deduced from the open mapping theorem which states that a continuous injective function which maps an open subset of $\mathbb{R}^n$ into $\mathbb{R}^n$ is open. In the case of a piecewise affine function, the latter result will be a side product of the following investigations.

We start with a study of injective piecewise affine functions. Our first observation is the fact that an injective piecewise affine function admits a collection of matrix-vector pairs, the matrices of which are nonsingular.

Proposition 2.3.2. *All matrices in the minimal collection of matrix-vector pairs corresponding to an injective piecewise affine function $f : \mathbb{R}^n \to \mathbb{R}^n$ are nonsingular.*

Proof. If $(A^1, b^1), \ldots, (A^k, b^k)$ is the minimal collection of matrix-vector pairs corresponding to f, then the sets $\sigma_i = \{x \in \mathbb{R}^n \mid f(x) = A^i x + b^i\}$ have nonempty interior. If $A^i v = 0$ for some nonvanishing vector $v \in \mathbb{R}^n$ and some index $i \in \{1, \ldots, k\}$, then every line $\{x_0 + \alpha v \mid \alpha \in \mathbb{R}\}$ will be mapped by the affine function $A^i x + b^i$ onto the point $A^i x_0 + b^i$. Choosing x_0 to be an interior point of σ_i, we thus obtain $f(x_0 + \alpha v) = f(x_0)$ for every sufficiently small α. Hence the equation $f(x) = f(x_0)$ has more than one solution, which contradicts the injectivity of f. Thus $A^i v = 0$ implies $v = 0$ which shows that A^i is nonsingular. $\qquad\square$

The following two-dimensional example demonstrates that the nonsingularity of the matrices is not sufficient for a piecewise affine function to be injective:

$$f(x, y) = \begin{cases} (x, y) & \text{if } (x, y) \in \sigma_1 = \{(x, y) \mid x \geq 0\} \\ (-x, y) & \text{if } (x, y) \in \sigma_2 = \{(x, y) \mid x \leq 0\} \end{cases}$$

What is the characteristic property that destroys injectivity in this case? Note that the collection $\{\sigma_1, \sigma_2\}$ is a polyhedral subdivision of $\mathbb{R}^2$ corresponding to f and that $L = \{(x, y) \in \mathbb{R}^2 \mid x = 0\}$ is the common face of both polyhedra in the partition. The subspace L divides $\mathbb{R}^2$ into two halfspaces, each halfspace containing one of the polyhedra. Since the matrices corresponding to both linear selection functions are nonsingular, the images $f(\sigma_1)$ and $f(\sigma_2)$ of the polyhedra σ_1 and σ_2 are two-dimensional polyhedra and the image of L is a common face of both polyhedra. The linear subspace $f(L)$ again divides the image space into two halfspaces. The injectivity is violated because the polyhedra $f(\sigma_1)$ and $f(\sigma_2)$ are contained in the same halfspace in the image space, while the polyhedra σ_1 and σ_2 are contained in different halfspaces in the preimage space. The following result shows that this situation typically destroys injectivity.

Proposition 2.3.3. *Let $f : \mathbb{R}^n \to \mathbb{R}^n$ be an injective piecewise affine function with polyhedral subdivision Σ. Whenever two polyhedra $\sigma_1, \sigma_2 \in \Sigma$ intersect in a common $(n-1)$-face, then so do $f(\sigma_1)$ and $f(\sigma_2)$.*

Proof. Suppose $(A^1, b^1), (A^2, b^2)$ are the matrix-vector pairs corresponding to σ_1 and σ_2, respectively, i.e., $f(x) = A^i x + b^i$ for every $x \in \sigma_i, i = 1, 2$. The continuity of f implies that both affine mappings coincide on $\sigma_1 \cap \sigma_2$. Since f is injective, Proposition 2.3.2 shows that the matrices A^1 and A^2 are nonsingular. Note that the nonsingular affine image of a convex polyhedron is again a convex polyhedron and faces are mapped onto faces of the same dimension. Hence the set $f(\sigma_1 \cap \sigma_2)$ is a common $(n-1)$-dimensional face of the polyhedra $f(\sigma_1)$ and $f(\sigma_2)$. If $f(\sigma_1)$ and $f(\sigma_2)$ do not intersect in the common $(n-1)$-face $f(\sigma_1 \cap \sigma_2)$, then there exists a common point y_0 which is not contained in the affine hull of the set $f(\sigma_1 \cap \sigma_2)$. If y_1 is a point in the relative interior of the common $(n-1)$-face $f(\sigma_1 \cap \sigma_2)$, then elementary geometric arguments show that any point in the relative interior of the line segment joining y_0 and y_1 is an interior point of both polyhedra $f(\sigma_1)$ and $f(\sigma_2)$. Clearly any point which is contained in the interior of both polyhedra $f(\sigma_1)$ and $f(\sigma_2)$ has two preimages, one of them being contained in the interior of σ_1 and the other one in the interior of σ_2. This, however, violates the injectivity assumption. Thus $f(\sigma_1)$ and $f(\sigma_2)$ intersect in the common $(n-1)$-face $f(\sigma_1 \cap \sigma_2)$. $\qquad\square$

A piecewise affine function $f : \mathbb{R}^n \to \mathbb{R}^m$ is said to be *coherently oriented on the polyhedron P* if there exists a polyhedral subdivision Σ of P with the properties

1. f coincides with an affine mapping on each polyhedron $\sigma \in \Sigma$
2. For every $\sigma \in \Sigma$ the dimensions of σ and $f(\sigma)$ coincide
3. Whenever two polyhedra σ and $\tilde{\sigma}$ intersect in a common $(\dim P - 1)$-face, then so do the polyhedra $f(\sigma)$ and $f(\tilde{\sigma})$

By abuse of language we call a piecewise affine function *coherently oriented* if it is coherently oriented on $\mathbb{R}^n$. Note that the dimensions of the preimage and image spaces of coherently oriented piecewise affine mappings coincide.

The next result is an immediate consequence of the latter definition and Proposition 2.3.3.

Proposition 2.3.4. *An injective piecewise affine function $f : \mathbb{R}^n \to \mathbb{R}^n$ is coherently oriented.*

Proof. Let Σ be a polyhedral subdivision of $\mathbb{R}^n$ and let $(A^1, b^1), \ldots, (A^k, b^k)$ be the minimal collection of matrix-vector pairs corresponding to f. By definition, f coincides with an affine selection function on each of the polyhedra $\sigma \in \Sigma$. Moreover, the matrices A^i are nonsingular in view of Proposition 2.3.2, and thus $\dim f(\sigma) = \dim \sigma = n$ for every $\sigma \in \Sigma$. Finally, if σ and $\tilde{\sigma}$ are two polyhedra in Σ which intersect in a common $(n-1)$-face, then Proposition 2.3.3 shows that $f(\sigma)$ and $f(\tilde{\sigma})$ also have this property. Hence f is coherently oriented. $\qquad\square$

The geometric condition presented in Proposition 2.3.3 can be equivalently formulated algebraically in terms of the determinants of the corresponding matrices A^1 and A^2.

Lemma 2.3.1. *If $\sigma_1, \sigma_2 \subseteq \mathbb{R}^n$ are n-dimensional convex polyhedra which intersect in a common $(n-1)$-face, A^1, A^2 are nonsingular $n \times n$-matrices, and b^1, b^2 are n-vectors such that*

$$A^1 x + b^1 = A^2 x + b^2 \text{ for every } x \in \sigma_1 \cap \sigma_2, \tag{2.24}$$

then $A^1(\sigma_1) + b^1$ and $A^2(\sigma_2) + b^2$ intersect in a common $(n-1)$-face if and only if $\det(A^1) \det(A^2) > 0$.

Proof. First we show that we may assume without loss of generality that

$$0 \in \sigma_1 \cap \sigma_2,$$
$$A^2 = I,$$
$$b^1 = 0,$$
$$b^2 = 0. \tag{2.25}$$

To check this, choose a vector $x_0 \in \sigma_1 \cap \sigma_2$, and define $A = (A^2)^{-1} A^1$. Elementary manipulations show that assumption (2.24) is equivalent to the identity $Ax = x$ for every $x \in (\sigma_1 - \{x_0\}) \cap (\sigma_2 - \{x_0\})$ and that the polyhedra $A^1(\sigma_1) + b^1$ and $A^2(\sigma_2) + b^2$ intersect in a common $(n-1)$-face if and only if this is true for the polyhedra $A(\sigma_1 - \{x_0\})$ and $\sigma_2 - \{x_0\}$. Moreover, $\det(A^1) \det(A^2) > 0$ if and only if $\det(A) > 0$. Hence, if the assumptions (2.25) do not hold, then we may replace σ_1, σ_2, A^1, and A^2 by $\sigma_1 - \{x_0\}$, $\sigma_2 - \{x_0\}$, A, and I, respectively. To further simplify the exposition, we next show that we may assume without loss of generality that

$$\sigma_1 \cap \sigma_2 \subseteq \{x \in \mathbb{R}^n | x_n = 0\}, \tag{2.26}$$

To see this, note that our assumptions (2.25) imply that $\sigma_1 \cap \sigma_2$ is contained in an $(n-1)$-dimensional linear subspace $L \subseteq \mathbb{R}^n$. Hence there exists an orthogonal matrix Q such that $QL = \{x \in \mathbb{R}^n | x_n = 0\}$. Moreover, σ_1 and σ_2 intersect in a common $(n-1)$-face if and only if this property holds for the polyhedra $Q\sigma_1$ and $Q\sigma_2$, and $A^1 x = x$ holds for every $x \in \sigma_1 \cap \sigma_2$ if and only if $Q A^1 Q^T y = y$ holds for every $y \in Q\sigma_1 \cap Q\sigma_2$. Since $A^1 \sigma_1$ and σ_2 intersect in a common $(n-1)$-face if and only if $(Q A^1 Q^T) Q\sigma_1$ and $Q\sigma_2$ have this property and since $\det(Q A^1 Q^T) = \det(A^1)$, we thus conclude that we may assume without loss of generality the validity of the assumptions (2.26), for otherwise, we replace σ_1, σ_2, and A^1 by $Q\sigma_1$, $Q\sigma_2$, and $Q A^1 Q^T$, respectively.

Defining $L = \{x \in \mathbb{R}^n | x_n = 0\}$, we deduce from (2.24) and (2.25) that

$$A^1 x = x \quad \text{for every } x \in L, \tag{2.27}$$

which shows that A^1 has the block structure

$$A^1 = \begin{pmatrix} I & 0 \\ 0 & \alpha \end{pmatrix}, \tag{2.28}$$

where α is a nonvanishing real number since A^1 is a nonsingular matrix. In view of (2.24) and (2.25), the set $A^1(\sigma_1) \cap \sigma_2$ is a common face of the polyhedra $A^1(\sigma_1)$ and σ_2. To prove the assertion, it thus suffices to show that the dimension of the intersection of $A^1(\sigma_1)$ and σ_2 is $(n-1)$ if and only if $\alpha > 0$. Defining $L^+ = \{x \in \mathbb{R}^n | x_n \geq 0\}$ and $L_- = \{x \in \mathbb{R}^n | x_n \leq 0\}$, the assumptions imply that one of the polyhedra, say σ_1, is contained in L_+, while σ_2 is contained in L_-. In view of the assumptions (2.24) and (2.25) the vector $x \in A^1(\sigma_1)$ whenever it is contained in $\sigma_1 \cap \sigma_2$. Hence the inclusion

$$\sigma_1 \cap \sigma_2 \subseteq A^1(\sigma_1) \cap \sigma_2 \tag{2.29}$$

holds and thus $\dim(A^1(\sigma_1) \cap \sigma_2) \geq (n-1)$, while $\dim(A^1(\sigma_1) \cap \sigma_2) = (n-1)$ if and only if the two polyhedra can be separated by a hyperplane. Since L is the affine hull of $\sigma_1 \cap \sigma_2$, the inclusion (2.29) shows that this separating hyperplane would necessarily be the linear subspace L, i.e., the intersection of $A^1(\sigma_1)$ and σ_2 has dimension $(n-1)$ if and only if $A^1(\sigma_1) \subseteq L_+$. In view of (2.27), $A^1(\sigma_1) \subseteq L_+$ if and only if L_+ is mapped by A^1 into L_+. Using (2.28), one readily verifies that $A^1(L_+) = L_+$ if and only if $\alpha > 0$ which holds if and only if $\det(A^1) > 0$. $\qquad\square$

The next proposition is sometimes used to define coherent orientation.

Proposition 2.3.5. *A piecewise affine function $f : \mathbb{R}^n \to \mathbb{R}^n$ with corresponding minimal collection of matrix-vector pairs $(A^1, b^1), \ldots, (A^k, b^k)$ is coherently oriented if and only if all matrices A^i have the same nonvanishing determinant sign.*

Proof. The "if"-part is an immediate consequence of Lemma 2.3.1 and the definition of coherent orientation. To see the "only if"-part, suppose f is coherently oriented and let Σ be a corresponding polyhedral subdivision of $\mathbb{R}^n$ which has the properties required by the definition of coherent orientation. In particular, $\dim f(\sigma) = \dim \sigma$, whence all matrices A^i have a nonvanishing determinant sign. Let $\tilde{\Sigma}$ be a subset of Σ such that the determinants of the matrices corresponding to the polyhedra in $\tilde{\Sigma}$ have positive determinant signs, while the remaining matrices have negative determinant signs. The assertion of the proposition is equivalent to the statement that either the collection $\tilde{\Sigma}$ or the collection $\Sigma \setminus \tilde{\Sigma}$ is empty. If both sets $\tilde{\Sigma}$ and $\Sigma \setminus \tilde{\Sigma}$ are nonempty, then both carriers $|\tilde{\Sigma}|$ and $|\Sigma \setminus \tilde{\Sigma}|$ have the same boundary which is a subset of the carrier $|\Sigma_{n-1}|$ of the $(n-1)$-skeleton of Σ. If $F \in \Sigma_{n-1}$ is a boundary face of $\tilde{\Sigma}$ and $\Sigma \setminus \tilde{\Sigma}$, then the subdivision property shows that there exist two polyhedra $\sigma \in \Sigma \setminus \tilde{\Sigma}$ and $\tilde{\sigma} \in \tilde{\Sigma}$ which intersect in the fact F. Hence by definition of coherent orientation the polyhedra $f(\sigma)$ and $f(\tilde{\sigma})$ intersect in a common $(n-1)$-face as well, and thus, in view of Lemma 2.3.1 the matrices of the corresponding affine functions have the same determinant sign. This, however, contradicts the fact that $\sigma \in \Sigma \setminus \tilde{\Sigma}$, while $\tilde{\sigma} \in \tilde{\Sigma}$ and thus proves that all matrices in the minimal collection of matrix-vector pairs corresponding to f have the same nonvanishing determinant sign. $\qquad\square$

The following example shows that a coherently oriented piecewise affine mapping is not necessarily injective.

Example 2.3.1. Consider the two-dimensional vectors

$$v_1 = (1,0), v_2 = (1,1), v_3 = (1,2), v_4 = (1,3), v_5 = (1,4), v_6 = (0,1),$$

$$w_1 = (1,0), w_2 = (0,1), w_3 = (-1,0), w_4 = (0,-1), w_5 = (1,0), w_6 = (0,1).$$

and define $f : \mathbb{R}^2 \to \mathbb{R}^2$ to be the piecewise linear function which coincides on cone$\{v_i, v_{i+1}\}$ with the linear mapping which carries v_i onto w_i and v_{i+1} onto w_{i+1}, $i = 1, \ldots, 5$, and which coincides with the identity outside of the union of the latter cones. It is easily verified that the determinants of all matrices are positive. Nevertheless every nonzero point in $\mathbb{R}^2$ has two preimages.

A closer examination of the latter mapping shows that, loosely spoken, the function f wraps $\mathbb{R}^2$ twice around the null vector. In particular, the mapping f is surjective which, in fact, is typical for coherently oriented piecewise affine functions.

Proposition 2.3.6. *A coherently oriented piecewise affine function is surjective.*

Proof. Consider a polyhedral subdivision $\Sigma = \{\sigma_1, \ldots, \sigma_k\}$ corresponding to a coherently oriented piecewise affine function $f : \mathbb{R}^n \to \mathbb{R}^n$. By Proposition 2.3.5, all matrices of the minimal collection of matrix-vector pairs corresponding to f are nonsingular, and hence all polyhedra $f(\sigma_i)$, $i = 1, \ldots, k$, are n-dimensional. Clearly f is surjective if and only if the union of all sets $f(\sigma_i)$, $i = 1, \ldots, k$, covers $\mathbb{R}^n$, i.e., if and only if this set has no boundary points. Suppose this is not the case. If the boundary of a finite union of n-dimensional polyhedra is nonempty, then obviously there exists a boundary point, say y_0 which is contained in the relative interior of some $(n-1)$-face F of one of the polyhedra, say $f(\sigma_1)$. Since all matrices corresponding to the affine selection functions of f are nonsingular, the face F is the image of an $(n-1)$-face of σ_1, which, due to the polyhedral subdivision property of the partition, is a common face with some other polyhedron, say σ_2. By definition of the coherent orientation, the polyhedra $f(\sigma_1)$ and $f(\sigma_2)$ intersect in the common face F and thus the relative interior point y_0 of F is an interior point of the union of the polyhedra $f(\sigma_1)$ and $f(\sigma_2)$. This, however, contradicts the fact the y_0 is a boundary point of the union of the polyhedra $f(\sigma_i)$, $i = 1, \ldots, k$. □

Another nice property of the function in the previous example is that it is open, i.e., it carries open sets onto open sets. Again, this is a typical property of a coherently oriented piecewise affine function. In fact, it is equivalent to the coherent orientation.

Proposition 2.3.7. *A piecewise affine function is coherently oriented if and only if it is open.*

Proof. A mere reformulation of the definition in terms of sequences shows that a function is open if and only if for every point x_0 and every sequence y_n converging to $y_0 = f(x_0)$ there exist a sequence x_n converging to x_0 such that $f(x_n) = y_n$ for every sufficiently large $n \in \mathbb{N}$. Using the B-derivative $f'(x_0; .)$ and the fact that the values of $f(x)$ and $f(x_0) + f'(x_0; x - x_0)$ coincide in a neighborhood of x_0, it suffices to find a sequence x_n such that $f'(x_0; x_n - x_0) = y_n - f(x_0)$ for sufficiently large $n \in \mathbb{N}$. By Propositions, 2.2.6 and 2.3.5 the B-derivative of a coherently oriented piecewise affine function is a coherently oriented piecewise linear function. It thus suffices to show that for a coherently oriented piecewise linear function $f : \mathbb{R}^n \to \mathbb{R}^n$ the origin in the image space is an interior point of the image of any neighborhood of zero. To see this, suppose y_n converges to zero. Proposition 2.3.6 yields the existence of a sequence x_n with $f(x_n) = y_n$. If $\{A^1, \ldots, A^k\}$ is a minimal set of matrices corresponding to f, then $x_n \in \{(A^i)^{-1} y_n \mid 1 \leq i \leq k\}$, and since y_n converges to zero, so does x_n. Hence the origin is an interior point of the image of every zero neighborhood, which proves that a coherently oriented piecewise affine function is open.

To prove the reverse statement, suppose the piecewise affine function $f : \mathbb{R}^n \to \mathbb{R}^n$ is open and let Σ be a polyhedral subdivision corresponding to f. If $\sigma \in \Sigma$ and $f(x) = Ax + b$ for every $x \in \sigma$, then the openness of f shows that A maps the interior of σ onto an open subset of $\mathbb{R}^n$ and hence $\det A \neq 0$. Thus $\dim f(\sigma) = \dim \sigma$ for every $\sigma \in \Sigma$. Next suppose σ and $\tilde{\sigma}$ intersect in a common $(n-1)$-face F and let x_0 be a relative interior point of F. Since the affine selection functions which are active on σ and $\tilde{\sigma}$, respectively, coincide on F, the set $f(F)$ is a common $(n-1)$-face of $f(\sigma)$ and $f(\tilde{\sigma})$. Hence any of the latter polyhedra is contained in one of the halfspaces induced by the hyperplane $\mathrm{aff}\, f(F)$. If the intersection of $f(\sigma)$ and $f(\tilde{\sigma})$ contains a point $y_1 \notin f(F)$, then both polyhedra $f(\sigma)$ and $f(\tilde{\sigma})$ are contained in the same halfspace, and thus $f(x_0)$ is a boundary point of the image $f(U)$ of every sufficiently small open neighborhood U of x_0. This, however, contradicts the openness of f and thus $f(\sigma)$ and $f(\tilde{\sigma})$ are contained in different halfspaces which shows $f(\sigma)$ and $f(\tilde{\sigma})$ intersect in the common $(n-1)$-face $f(F)$. Thus f is coherently oriented. $\qquad\qquad\square$

We have already seen in Example 2.3.1 that coherent orientation does not imply injectivity. The following slight modification of this example shows that surjectivity does not imply openness.

Example 2.3.2. Consider the following sets of vectors:

$$v_1 = (1, 0), v_2 = (1, 1), v_3 = (1, 2), v_4 = (0, 1),$$

$$w_1 = (1, 0), w_2 = (0, 1), w_3 = (1, 0), w_4 = (0, 1),$$

and let f be the piecewise linear function which coincides on the set $\mathrm{cone}\{v_i, v_{i+1}\}$ with the linear map that carries v_i onto w_i and v_{i+1} onto w_{i+1}, $i = 1, 2, 3$, and which coincides with the identity outside of the union of these cones. The matrices corresponding to $\mathrm{cone}\{v_1, v_2\}$ and to $\mathrm{cone}\{v_2, v_3\}$ are thus given by

$$A_1 = \begin{pmatrix} 1 & -1 \\ 0 & 1 \end{pmatrix} \quad \text{and} \quad A_2 = \begin{pmatrix} -1 & 1 \\ 2 & -1 \end{pmatrix}.$$

Since the determinants have different signs, the function is not coherently oriented and thus not open. Nevertheless, f is surjective.

For future reference, we summarize the statements of the Propositions 2.3.4–2.3.7 in the following theorem.

Theorem 2.3.1. *Let* $f : \mathbb{R}^n \to \mathbb{R}^n$ *be a piecewise affine function.*

1. If f is injective, then f is open.
2. If f is open, then f is surjective.
3. The function f is open if and only if it is coherently oriented.

The reverse implications of the first and second statement fail in general.

We conclude this section with a remarkable result on the number of preimages corresponding to a coherently oriented piecewise affine mapping. Recall that for a polyhedral subdivision Σ of $\mathbb{R}^n$ the set Σ_j denotes the j-skeleton of Σ, i.e., the set of all j-faces of the subdivision Σ.

Proposition 2.3.8. *If f is a coherently oriented piecewise affine function with corresponding polyhedral subdivision Σ, then any two points not contained in $f(|\Sigma_{n-2}|)$ have the same number of preimages. The number of preimages of a point contained in $f(|\Sigma_{n-2}|)$ does not exceed the number of preimages of points outside of this set.*

Proof. Since f is coherently oriented, every image vector has a finite number of preimages; at most one preimage corresponding to each function from the minimal collection of selection functions. Moreover, by Theorem 2.3.1, a coherently oriented piecewise affine function is open; hence the number of preimages cannot drop locally, i.e., if a vector y_0 has m preimages, then there exists a neighborhood of y_0 such that every vector in this neighborhood has at most m preimages. This is easily seen by choosing for each preimage an open neighborhood which does not contain any other preimage. The vector y_0 is an interior point of the intersection of the images of the chosen neighborhoods. Hence any vector sufficiently close to y_0 has a preimage in each of the m disjoint neighborhoods. What are the candidates in the image space where the number of preimages increases? Note that if y_n tends to y_0 and $f(x_n) = y_n$, then the nonsingularity of the matrices yields that x_n is bounded and since f is continuous, every cluster point of x_n will be a preimage of y_0. Hence a small perturbation of y_0 cannot cause a sudden occurrence of a new preimage somewhere far away from the preimages of y_0; it can only cause a bifurcation, i.e., some of the preimages of y_0 split into several branches if y_0 is perturbed. Such a bifurcation is certainly avoided if the function f is a local homeomorphism at each preimage of y_0. Clearly a coherently oriented piecewise affine function f is a local homeomorphism at every point $x_0 \in \text{int} \sigma_i$, where σ_i is a set from a partition corresponding to f. Moreover, if x_0 is contained in the relative interior of an $(n-1)$-face of a polyhedron from a polyhedral subdivision, then f has locally only two active affine selection functions. Lemma 2.3.1 shows that in this case f is a local homeomorphism if and only if the determinants of the corresponding matrices have the same nonvanishing sign. Hence if y_0 is not contained in $f(|\Sigma_{n-2}|)$, then no bifurcation occurs, i.e., the number of preimages cannot increase locally around y_0. Note that the set $f(|\Sigma_{n-2}|)$ is contained in the union of a finite number of $(n-2)$-dimensional affine subspaces. It is a matter of elementary geometrical insight to check that any two points not contained in

$f(|\Sigma_{n-2}|)$ can be joined by a polyhedral path which does not intersect $f(|\Sigma_{n-2}|)$. This shows that both points have the same number of preimages and thus proves the first part of the statement. The second part is easily checked since for every point in $f(|\Sigma_{n-2}|)$, one may find arbitrarily close to this point some point which is not contained in $f(|\Sigma_{n-2}|)$. As argued before, it is a consequence of the openness of f that the number of solutions cannot drop locally; hence any point in $f(|\Sigma_{n-2}|)$ has at most as many preimages as the points which are not contained in $f(|\Sigma_{n-2}|)$. $\square$

The latter result reveals a striking similarity between affine functions and coherently oriented piecewise affine functions. If an affine equation $Ax + b = y$ has a single solution for some $y = y_0$, then it has a single solution for all y, i.e., we can choose any test candidate y_0 to establish the injectivity of the mapping. Almost the same is true for coherently oriented piecewise affine functions, only that we have to choose y_0 carefully, namely outside of the image of $|\Sigma_{n-2}|$. This set, however, is a nullset and, moreover, its complement is open and dense. So from a probabilistic as well as from a topological point of view almost all points y_0 are good test points for the injectivity of the piecewise affine function.

Since $f(|\Sigma_{n-2}|)$ is a finite collection of lower dimensional polyhedra, every open set in the image space contains at least one point which is not contained in $f(|\Sigma_{n-2}|)$. We thus obtain the following corollary to the latter proposition which will be used frequently in the sequel.

Corollary 2.3.1. *Let $U \subseteq \mathbb{R}^n$ be an arbitrary open set. If $f : \mathbb{R}^n \to \mathbb{R}^n$ is coherently oriented and the equation $f(x) = y$ has a single solution for all $y \in U$, then f is a homeomorphism.*

2.3.2 Piecewise Affine Local Homeomorphisms

One readily verifies that a homeomorphism $f : \mathbb{R}^n \to \mathbb{R}^n$ is a local homeomorphism at every point $x \in \mathbb{R}^n$. The elementary example $f(x) = e^x$ shows that the reverse statement is generally false. In the affine case, however, the reverse statement does hold, i.e., an affine function is a homeomorphism whenever it is a local homeomorphism at some point $x \in \mathbb{R}^n$. The elementary example of the modulus function shows that the local homeomorphism property of a piecewise affine function at a single point does not provide any global information. However, the main results of the present section show that a piecewise affine function is a homeomorphism if it is a local homeomorphism at a finite number of distinguished points.

We begin with an elementary observation.

Proposition 2.3.9. *1. A local homeomorphism is an open mapping.*
2. A piecewise affine local homeomorphism $f : \mathbb{R}^n \to \mathbb{R}^n$ is coherently oriented.

Proof. To prove the first statement, we have to show that for every sequence of image points $y^k \in Y$ converging to an image point $y^0 \in Y$ there exists a sequence of preimages $x^k \in X$ converging to some point $x^0 \in X$ such that $f(x^0) = y^0$ and $f(x^k) = y^k$ for every sufficiently large $k \in \mathbb{N}$. To see this, choose an arbitrary element $x^0 \in f^{-1}(\{y^0\})$. Since f is a local homeomorphism at x^0, there exist neighborhoods $U \subseteq X$ of x^0 and $V \subseteq Y$ of y^0 such that f maps U homeomorphically onto V. Since the sequence y^k will eventually be captured in the neighborhood V of y^0, we can find for every sufficiently large $k \in \mathbb{N}$ a unique vector $x^k \in U$ with $f(x^k) = y^k$, which proves the assertion. The second statement is an immediate consequence of part 1 and Theorem 2.3.1. □

The following proposition shows that the local homeomorphism property for a piecewise affine function is passed on from lower dimensional faces to higher dimensional faces of a corresponding polyhedral subdivision.

Proposition 2.3.10. *Let* $f : \mathbb{R}^n \to \mathbb{R}^n$ *be a piecewise affine function and* Σ *be a corresponding subdivision of* $\mathbb{R}^n$. *Suppose* $\emptyset \neq \Sigma_k = \{F_1, \ldots, F_r\}$ *and let* $x_i \in \operatorname{relint} F_i$, $i = 1, \ldots, r$. *If* f *is a local homeomorphism at every point* x_i, $i = 1, \ldots, r$, *then* f *is a local homeomorphism at every point* $x_0 \notin |\Sigma_{k-1}|$.

Proof. In view of part 4 of Proposition 2.2.6, the piecewise linear functions $f'(x_i; .)$, $i \in \{1, \ldots, r\}$ are local homeomorphisms at the origin. Since the latter functions are positively homogeneous, they are global homeomorphisms and thus, using again Proposition 2.2.6, we conclude that the function f is a local homeomorphism at every interior point of $|\Sigma(x_i)|$, $i = 1, \ldots, r$. Given a point $x_0 \notin |\Sigma_{k-1}|$, it thus suffices to prove that there exists a vector x_i such that x_0 is an interior point of $|\Sigma(x_i)|$.

To see this, note first that x_0 is an interior point of $|\Sigma(x_0)|$. Moreover, every polyhedron $\sigma_i \in \Sigma(x_0)$ has a unique face G_i containing x_0 as a relative interior point. If G_j is the face corresponding to another polyhedron $\sigma_j \in \Sigma(x_0)$, then the faces G_i and G_j are both faces of the common face of the polyhedra σ_i and σ_j. Since G_i as well as G_j contain x_0 as a relative interior point, they thus coincide. Hence we conclude that there exists a unique face G which is a common face of all polyhedra $\sigma \in \Sigma(x_0)$ and contains x_0 as a relative interior point.

Since $x_0 \notin |\Sigma_{k-1}|$, the dimension of G is at least k and thus there exists a k-face F_{i_0} which is a face of G. Since G is a face of σ for every $\sigma \in \Sigma(x_0)$, the same holds for the face F_{i_0}. Since by assumption x_{i_0} is a relative interior point of F_{i_0}, we thus conclude that $x_{i_0} \in \sigma$ for every $\sigma \in \Sigma(x_0)$. Hence $\Sigma(x_0) \subseteq \Sigma(x_{i_0})$ and since x_0 is an interior point of $|\Sigma(x_0)|$, it is also an interior point of $|\Sigma(x_{i_0})|$, which completes the proof. □

Choosing k to be the dimension of the lineality space of Σ, the latter proposition implies that it suffices to check the local homeomorphism property at a finite number of distinguished points.

During the proof of Proposition 2.3.8 we have already established the next result.

Proposition 2.3.11. *A coherently oriented piecewise affine function* $f : \mathbb{R}^n \to \mathbb{R}^n$ *with corresponding polyhedral subdivision* Σ *of* $\mathbb{R}^n$ *is a local homeomorphism at every point* $x \in \mathbb{R}^n \setminus |\Sigma_{n-2}|$.

The following theorem contains the main results of this section. While the importance of the first statement is clear, the usefulness of the second assertion will only
become apparent in subsequent sections.

Theorem 2.3.2. *1. A piecewise affine local homeomorphisms $f : \mathbb{R}^n \to \mathbb{R}^n$ is a
homeomorphism.*

*2. Let $n \geq 3$, $f : \mathbb{R}^n \to \mathbb{R}^n$ be a piecewise linear function, and $v_1, \ldots, v_k \in \mathbb{R}^n$
be unit generators of the extremal rays of a corresponding pointed conical
subdivision. If f is a local homeomorphism at every point v_i, $i = 1, \ldots, k$,
then f is a homeomorphism.*

To prove the latter theorem, one has to employ quite general results, which belong
to the realm of algebraic topology. In fact, a proof is fairly easy if one uses
Browder's Theorem on local homeomorphisms which are covering maps and the
homotopy lifting property for covering maps. It is beyond the scope of this section
to provide proofs of these classical theorems. Nevertheless, we will introduce the
necessary topological notions to formulate them and show how they can be used to
prove Theorem 2.3.2. The reader who is willing to accept the validity of the latter
statement may skip the rest of this section. For references concerning the following
results we refer to Sect. 2.3.5 below

In order not to overload the exposition with unnecessary topological details, we
restrict ourselves to subsets X of $\mathbb{R}^n$, the topology on X being induced by the
Euclidean metric. Note that a closed subset of X is not necessarily a closed subset
of $\mathbb{R}^n$ unless X itself is a closed subset of $\mathbb{R}^n$. In fact, a set $A \subseteq X$ is a closed subset
of X if the limit point of every convergent sequence in A belongs to A, provided the
limit point is contained in X. A mapping $f : X \to Y$ is called *closed* if the image
$f(A)$ of every closed subset $A \subseteq X$ is a closed subset of Y.

We begin with a notion which is slightly stronger than the notion of a local
homeomorphism. A continuous mapping $f : X \to Y$ is called a *covering map*
if every $y \in Y$ admits an open neighborhood $V \subseteq Y$ such that $f^{-1}(V)$ is the
disjoint union of open sets, each of which is mapped homeomorphically onto V by
f. A typical example of a covering map is the mapping $f : \mathbb{R} \to S^1$ given by
$f(t) = (\sin t, \cos t)$, where S^1 denotes the unit sphere in $\mathbb{R}^2$. One easily verifies
that a covering map is a local homeomorphism. The reverse statement, however, is
false in general. The classical example is the restriction of the above function f to
the interval $(0, 4\pi)$. In fact, the point $(1, 0)$ does not admit an open neighborhood
$V \subseteq (0, 4\pi)$, the preimage of which is the disjoint union of open sets, each of
which is mapped homeomorphically onto V by f. A covering map $f : X \to Y$ is
called *finite* if each vector $y \in Y$ has a finite number of preimages. It is well known
that a local homeomorphism $f : X \to Y$ is a finite covering map if the number
of preimages of every point $y \in Y$ is finite and constant. In fact, if $y \in Y$ and
$f^{-1}(y) = \{x_1, \ldots, x_l\}$, then choose disjoint open neighborhoods $V_i \subseteq X$ of x_i
which are mapped homeomorphically onto the open sets $f(V_i)$ and let $U = \cap_{i=1}^l$
$f(V_i)$. Defining $\tilde{V}_i = f_i^{-1}(U)$, where $f_i^{-1} : f(V_i) \to V_i$ is the local inverse of
f at x_i, one readily verifies that the sets $\tilde{V}_i$ are disjoint open sets, each of which

is mapped homeomorphically onto the open set U and that $f^{-1}(U) = \cup_{i=1}^{l} \tilde{V}_i$ since every point in U has precisely l preimages. Thus Proposition 2.3.8 shows that the restriction of a coherently oriented piecewise affine function f to the carrier $|\Sigma \backslash \Sigma_{n-2}|$ is a finite covering map. Another useful condition ensuring that a local homeomorphism is a finite covering map can be obtained by focusing attention to mappings with a connected image space. Recall that a subset $Y \subseteq \mathbb{R}^n$ is called *connected* if Y cannot be separated by two open sets, i.e., if Y is a subset of the union of two open sets $U, V \subseteq \mathbb{R}^n$, then $U \cap V \neq \emptyset$.

Theorem 2.3.3 (Browder's Theorem [6], Theorem 7). *Let* $X \subseteq \mathbb{R}^n$, $Y \subseteq \mathbb{R}^m$, *and* $f : X \to Y$ *be a local homeomorphism. If* f *is closed and* Y *is connected, then* f *is a finite covering map.*

A fundamental characteristic of covering maps is comprised in the so-called homotopy lifting property.

Theorem 2.3.4 (Homotopy Lifting Theorem [78], Chap. 2.2, Theorem 3). *Let* $X \subseteq \mathbb{R}^n$, $Y \subseteq \mathbb{R}^m$, *and* $S \subseteq \mathbb{R}^k$ *be sets and let* $f : X \to Y$, $p : S \to X$, *and* $H : S \times [0,1] \to Y$ *be continuous functions with* $H(s,0) = f(p(s))$ *for every* $s \in S$. *If* f *is a covering map, then there exists a continuous function* $\hat{H} : S \times [0,1] \to X$ *such that* $f \circ \hat{H} = H$ *and* $\hat{H}(s,0) = p(s)$ *for every* $s \in S$.

If S is connected, then the mapping $\hat{H}$ is unique as a consequence of the following theorem.

Theorem 2.3.5 ([78], Chap. 2.2, Theorem 2). *Let* $X \subseteq \mathbb{R}^n$, $Y \subseteq \mathbb{R}^m$, *and* $Z \subseteq \mathbb{R}^k$ *be sets,* $f : X \to Y$ *be a covering map, and let* $\hat{G}, \hat{H} : Z \to X$ *be two continuous mappings with* $f \circ \hat{G} = f \circ \hat{H}$. *If* Z *is connected and* $\hat{G}(z) = \hat{H}(z)$ *for some point* $z \in Z$, *then* $\hat{G} = \hat{H}$.

To formulate the main theorem, we need some additional notions. A set $X \subseteq \mathbb{R}^n$ is called *path connected* if for any two points $x_0, x_1 \in X$ there exists a continuous function $p : [0,1] \to X$ such that $p(0) = x_0$ and $p(1) = x_1$. The function p is called a *path* joining x_0 and x_1. Note that a path connected set is connected, but that the reverse statement fails in general. A *loop* is a path p with $p(0) = p(1)$. A set $X \subseteq \mathbb{R}^n$ is called *simply connected* if it is path connected and every loop can be continuously deformed into a point, i.e., for every path $p : [0,1] \to X$ with $p(0) = p(1) = x_0$ there exists a point $x_1 \in X$ and a continuous function $H : [0,1] \times [0,1] \to X$ such that $H(s,0) = p(s)$ and $H(s,1) = x_1$ for every $s \in [0,1]$ and $H(0,t) = H(1,t)$ for every $t \in [0,1]$. Since X is assumed to be path connected, we may choose any point $x_1 \in X$.

With the aid of Browder's Theorem and the Homotopy Lifting Theorem, one can prove the following result which relates local and global homeomorphisms.

Theorem 2.3.6. *Suppose* $X \subseteq \mathbb{R}^n$ *is path connected,* $Y \subseteq \mathbb{R}^m$ *is simply connected, and* $f : X \to Y$ *is a local homeomorphism. If* f *is closed, then* f *is a homeomorphism.*

Proof. By Proposition 2.3.9, a local homeomorphism is an open mapping. Moreover, since by assumption f is closed, the set $f(X)$ is an open and closed subset of Y and hence coincides with Y which establishes the surjectivity of f. It thus remains to show that f is injective. Suppose $x_0, x_1 \in X$ are two points with $y_0 = f(x_0) = f(x_1)$, and consider a path $p : [0, 1] \to X$ joining x_0 and x_1. Since Y is simply connected, there exists a function $H : [0, 1] \times [0, 1] \to Y$ such that $H(s, 0) = f(p(s))$ and $H(s, 1) = y_0$ for every $s \in [0, 1]$, and $H(0, t) = H(1, t)$ for every $t \in [0, 1]$. By Theorem 2.3.3 the function f is a finite covering map and hence the Homotopy Lifting Theorem shows that there exists a continuous function $\hat{H} : [0, 1] \times [0, 1] \to X$ such that

$$f(\hat{H}(s, t)) = H(s, t) \quad \text{for every } (s, t) \in [0, 1] \times [0, 1], \tag{2.30}$$

$$\hat{H}(s, 0) = p(s) \quad \text{for every } s \in [0, 1]. \tag{2.31}$$

Equation (2.30) shows that $\hat{H}([0, 1], 1) \subseteq f^{-1}(H([0, 1], 1)) = f^{-1}(y_0)$. Since f is a finite covering map, the latter set is finite and since the set $[0, 1] \times \{1\}$ is connected and $\hat{H}$ is continuous, we deduce that the set $\hat{H}([0, 1], 1)$ is connected as well and thus consists of a single point. This implies that

$$\hat{H}(0, 1) = \hat{H}(1, 1). \tag{2.32}$$

Since by assumption $H(0, t) = H(1, t)$, we can use Theorem 2.3.5 to deduce from (2.30), and (2.32) that the functions $\hat{H}(0, .)$ and $\hat{H}(1, .)$ coincide. Thus (2.31) yields

$$p(0) = \hat{H}(0, 0) = \hat{H}(1, 0) = p(1),$$

which proves that $x_0 = x_1$ and thus shows that f is injective.   □

Proof of Theorem 2.3.2

1. Since $\mathbb{R}^n$ is simply connected, it suffices to prove that a piecewise affine function is closed. This is easily seen since a linear function maps closed sets onto closed sets; hence the image of a closed set by a piecewise affine function is the finite union of closed sets and thus closed.
2. In view of Theorem 2.3.1 it suffices to prove that f is injective. Setting $k = 1$, we can deduce from Proposition 2.3.10 that f is a local homeomorphism at every point $x \neq 0$. If $A^1, \ldots, A^k$ is a minimal collection of matrices corresponding to the piecewise linear function f, then the latter property shows that every matrix A^i is nonsingular. Hence the equation $f(x) = 0$ has the unique solution $x = 0$ and it thus suffices to prove that the function $g : \mathbb{R}^n \backslash \{0\} \to \mathbb{R}^n \backslash \{0\}$ defined by $g(x) = f(x)$ is injective. As argued above, the mapping g is a local homeomorphism. Since $n \geq 3$, the set $\mathbb{R}^n \backslash \{0\}$ is simply connected. In view of Theorem 2.3.6, it thus suffices to prove that g is closed. Note that C is a closed

subset of $\mathbb{R}^n \backslash \{0\}$ if and only if $C \cup \{0\}$ is a closed subset of $\mathbb{R}^n$. Since $f(x) = 0$ holds if and only if $x = 0$, we deduce that $f(\hat{C} \backslash \{0\}) = f(\hat{C}) \backslash \{0\}$ for every subset $\hat{C} \subseteq \mathbb{R}^n$. We have already argued in the proof of the first part of the theorem that a piecewise affine mapping is a closed mapping. Hence if $\hat{C}$ is a closed subset of $\mathbb{R}^n$, so is the set $f(\hat{C})$. Thus $g(C) = g(\hat{C} \backslash \{0\})$ is closed for every closed subset $C = \hat{C} \backslash \{0\}$ of $\mathbb{R}^n \backslash \{0\}$. □

2.3.3 The Factorization Lemma

We have seen in the last section that a piecewise affine function is a homeomorphism if and only if it is a local homeomorphism at a finite number of distinguished points. It has also been pointed out before that a piecewise affine function $f : \mathbb{R}^n \to \mathbb{R}^n$ is a local homeomorphism at $x \in \mathbb{R}^n$ if and only if its B-derivative $f'(x; .)$ is a homeomorphism. This reduces the homeomorphism problem for piecewise affine functions to the same problem for piecewise linear functions. In view of part 2 of Theorem 2.3.2 and the fact that every piecewise linear function admits a corresponding pointed conical subdivision, the homeomorphism problem for a piecewise linear function is reduced to the homeomorphism problem for a finite number of B-derivatives. At first glance this does not seem to be a reduction of the problem. However, note that the B-derivative $f'(x; .)$ is taken at nonvanishing points x. If Σ is a subdivision corresponding to the piecewise linear function f and $x \neq 0$, then $\mathrm{lin}\{x\}$ is a subset the lineality space of the localization $\Sigma'(x)$ which is a subdivision corresponding to $f'(x; .)$. So we have actually reduced the homeomorphism problem for a piecewise linear function to the homeomorphism problem for a finite number of piecewise linear functions which admit a corresponding subdivision with nonvanishing lineality space. In order to exploit this property, we would like to factor out the lineality space. It will be shown below that this is indeed possible. We begin with an elementary observation.

Lemma 2.3.2. *If $f : \mathbb{R}^n \to \mathbb{R}^n$ is a piecewise linear function, Σ is a corresponding conical subdivision of $\mathbb{R}^n$, and L is a linear subspace of the lineality space of Σ, then f is linear on L and $f(v + w) = f(v) + f(w)$ for every $v \in L$ and $w \in L^{\perp}$.*

Proof. Since L is a subspace of the lineality space of every cone $\sigma \in \Sigma$, L is a subset of every cone $\sigma \in \Sigma$ which shows that all linear selection functions of f coincide on L and thus f is linear on L. Moreover, the decomposition theorem for polyhedra yields the identity $\sigma = L + \sigma \cap L^{\perp}$. Hence, if $v + w \in \sigma$, then $w \in \sigma \cap L^{\perp}$, while $v \in L \subseteq \sigma$. Since f coincides with a linear function on σ, we thus obtain $f(v + w) = f(v) + f(w)$. □

Let us denote by Π_L the *orthogonal projection* onto a linear subspace $L \subseteq \mathbb{R}^n$. Recall that the orthogonal projection onto a linear subspace is a linear function. If $f : \mathbb{R}^n \to \mathbb{R}^n$ is a piecewise linear function and L is a linear subspace of the lineality space of a corresponding conical subdivision of $\mathbb{R}^n$, then the latter lemma states that f is a linear function on L and hence $f(L)$ is a linear subspace of $\mathbb{R}^n$. We may thus define the function $f_{L^{\perp}} : \mathbb{R}^n \to \mathbb{R}^n$ by

$$f_{L^{\perp}}(x) = \Pi_{f(L)^{\perp}}(f(x)). \tag{2.33}$$

The piecewise linear function $f_{L^{\perp}}$ if called the *factor* of f with respect to $L^{\perp}$. Recall that a linear function f is a homeomorphism if and only if for some linear subspace L the function maps L and its orthogonal complement $L^{\perp}$ homeomorphically onto $f(L)$ and $f(L^{\perp})$, respectively and, in addition, the linear subspaces $f(L)$ and $f(L^{\perp})$ intersect only at the origin. One readily verifies that the latter condition is equivalent to the requirement that f maps L homeomorphically onto $f(L)$ and $f_{L^{\perp}}$ maps $L^{\perp}$ homeomorphically onto $f(L)^{\perp}$. The next result provides a generalization of this fact to piecewise linear functions.

Lemma 2.3.3 (Factorization Lemma). *If* $f : \mathbb{R}^n \to \mathbb{R}^n$ *is a piecewise linear function,* Σ *a corresponding conical subdivision of* $\mathbb{R}^n$, *and* L *a linear subspace of the lineality space of* Σ, *then* f *is a homeomorphism if and only if* f *maps* L *homeomorphically onto* $f(L)$ *and the factor* $f_{L^{\perp}}$ *maps* $L^{\perp}$ *homeomorphically onto* $f(L)^{\perp}$.

Proof. Clearly, if f is a homeomorphism, then it maps L homeomorphically onto $f(L)$. We may thus assume that f has the latter property and show that f is a homeomorphism if and only if $f_{L^{\perp}}$ maps $L^{\perp}$ homeomorphically onto $f(L)^{\perp}$. To simplify the exposition, we define $g : L^{\perp} \to f(L)^{\perp}$ by $g(x) = f_{L^{\perp}}(x)$. Clearly if f maps L homeomorphically onto $f(L)$, then the linear subspaces $L^{\perp}$ and $f(L)^{\perp}$ have the same dimension, say m. Hence, if N is an $m \times n$-matrix with $N f(L)^{\perp} = \mathbb{R}^m$ and M is an $n \times m$-matrix with $M \mathbb{R}^m = L$, then the function $h = N \circ g \circ M : \mathbb{R}^m \to \mathbb{R}^m$ is a piecewise linear mapping and g is a homeomorphism if and only if h is a homeomorphism. In view of Theorem 2.3.1, the assertion of the lemma is thus equivalent to the statement that the homeomorphism property of any of the functions f or g implies the injectivity of the other one.

To prove this, we consider the equation $f(x) = y$, and let $v = \Pi_L(x)$ and $w = \Pi_{L^{\perp}}(x)$. The equation $f(x) = y$ is thus equivalent to the two equations

$$\Pi_{f(L)}(f(v + w)) = \Pi_{f(L)}(y),$$
$$\Pi_{f(L)^{\perp}}(f(v + w)) = \Pi_{f(L)^{\perp}}(y). \tag{2.34}$$

Lemma 2.3.2 yields

$$f(v + w) = f(v) + f(w). \tag{2.35}$$

Since $f(v) \in f(L)$, the identity $\Pi_{f(L)^{\perp}}(f(v)) = 0$ holds, and, by definition of the function g, $g(w) = \Pi_{f(L)^{\perp}}(f(w))$. Thus, in view of (2.35) and the linearity of the orthogonal projection equations (2.34) are equivalent to

$$f(v) = \Pi_{f(L)}(y - f(w)),$$

$$g(w) = \Pi_{f(L)^\perp}(y). \tag{2.36}$$

If, on the one hand, g is a homeomorphism, then w is uniquely determined by the second equation, and since by assumption f maps L homeomorphically onto $f(L)$, v is uniquely determined by the first equation; hence $f(x) = y$ has a unique solution and thus f is injective. If, on the other hand, f is a homeomorphism and y is an arbitrary point in $f(L)^\perp$, then the equation $f(x) = y$ has a unique solution x which can be uniquely represented as $x = v + w$ with $v \in L$ and $w \in L^\perp$. Since $f(v + w) = y$ if and only if v and w satisfy equation (2.36), and since $y \in f(L)^\perp$, we conclude that $g(w) = y$ has a unique solution and thus g is injective. $\qquad \square$

We close this section with a collection of some useful results concerning the factors of piecewise linear functions.

Proposition 2.3.12. *Let* $f : \mathbb{R}^n \to \mathbb{R}^n$ *be a piecewise linear function,* Σ *be a corresponding conical subdivision, let L be a linear subspace of the lineality space of Σ, and let $f_{L\perp}$ be the factor of f with respect to $L^\perp$.*

1. *The collection* $\tilde{\Sigma} = \{\sigma \cap L^\perp | \sigma \in \Sigma\}$ *is a conical subdivision of $L^\perp$.*
2. *The factor $f_{L\perp}$ coincides with an affine function on each cone $\sigma \cap L^\perp$.*
3. *If f is coherently oriented on $\mathbb{R}^n$, then its factor $f_{L\perp}$ is coherently oriented on $L^\perp$.*

Proof. 1. Since L is a subset of the lineality space of σ, we may decompose σ as

$$\sigma = L + \sigma \cap L^\perp. \tag{2.37}$$

Hence Lemma 2.1.2 shows that F is a face of σ if and only if $F = L + F \cap L^\perp$ and $F \cap L^\perp$ is a face of $\sigma \cap L^\perp$. Moreover, since L and $L^\perp$ are mutually orthogonal, we obtain

$$\dim F = \dim L + \dim F \cap L^\perp. \tag{2.38}$$

To see that the nonempty intersection of two cones in $\tilde{\Sigma}$ is a common proper face of both cones, note that, 2.37 implies that $\sigma \cap \tilde{\sigma} = L + (\sigma \cap L^\perp) \cap (\tilde{\sigma} \cap L^\perp)$ for every $\sigma, \tilde{\sigma} \in \Sigma$. Hence if F is the common proper face of σ and $\tilde{\sigma}$, then the cones $\sigma \cap L^\perp$ and $\tilde{\sigma} \cap L^\perp$ intersect in the face $F \cap L^\perp$ which is proper in view of (2.38). Moreover, since $\dim \sigma = n$ for every $\sigma \in \Sigma$, we deduce from (2.37) that $\dim \sigma \cap L^\perp = \dim L^\perp$. Finally, the collection of all polyhedra $\sigma \cap L^\perp, \sigma \in \Sigma$, certainly covers $L^\perp$ and thus forms a polyhedral subdivision of $L^\perp$.
2. This assertion is trivial since $\sigma \cap L^\perp \subseteq \sigma$ and $f_{L\perp}$ is linear on each cone $\sigma \in \Sigma$.
3. We have already seen above that the function $f_{L\perp}$ coincides with an affine function on every cone $\sigma \cap L^\perp$ and that the collection of all cones $\sigma \cap L^\perp$, $\sigma \in \Sigma$, is a conical subdivision of $L^\perp$. To prove that $f_{L\perp}$ is coherently oriented on $L^\perp$, we first show that for every $\sigma \in \Sigma$ the decomposition

$$f(\sigma) = f(L) + f_{L\perp}(\sigma \cap L^\perp) \tag{2.39}$$

holds. Note that Lemma 2.3.2 yields the identity $f(v + w) = f(v) + f(w)$ for every $v \in L$ and $w \in L^{\perp}$. Hence (2.37) implies that

$$f(\sigma) = f(L) + f(\sigma \cap L^{\perp}). \tag{2.40}$$

In particular, $f(L)$ is a subspace of the lineality space of $f(\sigma)$. Hence the decomposition theorem for polyhedra shows that $f(\sigma) = f(L) + f(\sigma) \cap f(L)^{\perp}$ and a fortiori $\Pi_{f(L)^{\perp}}(f(\sigma)) = f(\sigma) \cap f(L)^{\perp}$. Combining the latter identities, we thus obtain $f(\sigma) = f(L) + \Pi_{f(L)^{\perp}}(f(\sigma))$. Since (2.40) implies that $\Pi_{f(L)^{\perp}}(f(\sigma)) = \Pi_{f(L)^{\perp}}(f(\sigma \cap L^{\perp}))$, we thus conclude that

$$f(\sigma) = f(L) + \Pi_{f(L)^{\perp}}\left(f\left(\sigma \cap L^{\perp}\right)\right), \tag{2.41}$$

which in view of the definition (2.33) of $f_{L^{\perp}}$ yields (2.39).

Since f is coherently oriented, the polyhedra $f(\sigma)$ are n-dimensional and thus (2.39) shows that $\dim f_{L^{\perp}}(\sigma \cap L^{\perp}) = \dim L^{\perp}$. Moreover, in view of the first statement of the present proposition the two polyhedra σ and $\tilde{\sigma}$ intersect in a face of dimension $(n-1)$ if and only if the polyhedra $\sigma \cap L^{\perp}$ and $\tilde{\sigma} \cap L^{\perp}$ intersect in a face of dimension $(\dim L^{\perp} - 1)$. Similarly, we deduce from (2.39) that the polyhedra $f(\sigma)$ and $f(\tilde{\sigma})$ intersect in a face of dimension $(n - 1)$ if and only if the polyhedra $f_{L^{\perp}}(\sigma \cap L^{\perp})$ and $f_{L^{\perp}}(\tilde{\sigma} \cap L^{\perp})$ intersect in a face of dimension $(\dim f(L) - 1)$. Since f is coherently oriented, the dimensions of L and $f(L)$ coincide. Hence we conclude that the polyhedra $f_{L^{\perp}}(\sigma \cap L^{\perp})$ and $f_{L^{\perp}}(\tilde{\sigma} \cap L^{\perp})$ intersect in a face of dimension $(\dim L^{\perp} - 1)$ whenever the polyhedra $\sigma \cap L^{\perp}$ and $\tilde{\sigma} \cap L^{\perp}$ have this property. Thus $f_{L^{\perp}}$ is indeed coherently oriented on $L^{\perp}$.
□

2.3.4 The Branching Number Theorem

For practical purposes, the most important necessary condition for a piecewise affine function to be a homeomorphism is its coherent orientation. Given a collection of matrix-vector pairs corresponding to a piecewise affine function, this condition is easily checked. We have already given an example which shows that coherent orientation is not sufficient for a piecewise affine function to be a homeomorphism. One might now try to develop criteria which are necessary and sufficient. However, necessary and sufficient criteria are nothing else but restatements of the definition and, although shedding light on the problem, are usually not very helpful in practice. It is much more appropriate to look for sufficient conditions which are on the one hand applicable to a large class of piecewise affine functions and, on the other hand, are relatively easy to verify. There are of course two structures on which one can impose additional assumptions; the collection of matrix-vector pairs on the one hand, and the polyhedral subdivision corresponding to a piecewise affine function, on the other hand. Note that both approaches are not independent since admissible subdivision structures contain information about admissible matrix-vector pairs and vice versa. The subdivision structure is, however, more geometric in spirit and thus corresponding conditions are easier to envision. Moreover, it turns out that the condition which we discuss below can be nicely applied to some important examples of piecewise affine functions. This is the main reason to confine our presentation to sufficient conditions which rely on the subdivision structure.

It turns out that a decisive property of a polyhedral subdivision in the context of sufficient homeomorphism conditions is the maximal number of full-dimensional polyhedra containing a face of a given dimension. We therefore introduce the *k-th branching number* of a polyhedral subdivision Σ of a polyhedron $P \subseteq \mathbb{R}^n$ as the maximal number of polyhedra in Σ containing a face of dimension (dim $P - k$), where $k \in \{1, \dots, \dim P - l\}$ and l is the dimension of the lineality space of Σ. One readily verifies that the first branching number of Σ is 2, provided that Σ does not consist of the set P alone. The following lemma shows that the branching numbers do not grow if one passes from a polyhedral subdivision Σ of $\mathbb{R}^n$ corresponding to a piecewise affine function f to the localization' $\Sigma'(x)$ corresponding to the B-derivative $f'(x; .)$, or, in the case of a conical subdivision and a piecewise linear function, to the subdivision corresponding to a factor of f (cf. Proposition 2.3.12).

Lemma 2.3.4. *Let Σ be a polyhedra subdivision of $\mathbb{R}^n$ and let $k \in \{2, \dots, n - l\}$, where l is the dimension of the lineality space of Σ.*

1. *If $x \in \mathbb{R}^n$, then the k-th branching number of the localization $\Sigma'(x)$ does not exceed the k-th branching number of Σ.*
2. *If Σ is a conical subdivision of $\mathbb{R}^n$ and L is a subset of the lineality space of Σ, then the k-th branching number of the subdivision $\tilde{\Sigma} = \{\sigma \cap L^{\perp} | \sigma \in \Sigma\}$ of $L^{\perp}$ does not exceed the k-th branching number of Σ.*

Proof. 1. Recall that $\Sigma'(x) = \{\text{cone}(\sigma - \{x\}) | \sigma \in \Sigma, x \in \sigma\}$. The assertion is an immediate consequence of the fact that a subset $F' \subseteq \mathbb{R}^n$ is a face of $\text{cone}(\sigma - \{x\})$ if and only if $F' = \text{cone}(F - \{x\})$, where F is a face of σ with $x \in F$. Clearly the dimensions of F' and F both coincide with the dimension of $\text{lin}(F - \{x\})$ and hence if a face F' of dimension $(n - k)$ is contained in $\text{cone}(\sigma - \{x\})$, then the corresponding face F has also dimension $(n - k)$ and is contained in the polyhedron σ.

2. To see the second assertion, note that a set $F \subseteq \mathbb{R}^n$ is a face of dimension $(n - k)$ of a cone $\sigma \in \Sigma$ if and only if $F \cap L^{\perp}$ is a face of dimension $(\dim L^{\perp} - k)$ of the cone $\sigma \cap L^{\perp}$. $\square$

The following theorem is the main result of this section. It shows how the branching numbers can be used to decide whether a piecewise affine function is a homeomorphism.

Theorem 2.3.7. *Let $f : \mathbb{R}^n \to \mathbb{R}^n$ be a piecewise affine function with corresponding polyhedral subdivision Σ and let l be the dimension of the lineality space of Σ.*

If either $l = (n - 1)$ or there exists a number $k \in \{2, \ldots, n - l\}$ such that the k-th branching number does not exceed $2k$, then f is a homeomorphism if and only if it is coherently oriented.

Proof. Proposition 2.3.11 states that a coherently oriented piecewise affine function is a local homeomorphism at every point in $\mathbb{R}^n \setminus \Sigma_{n-2}$. Hence if the dimension of the lineality space of Σ is $n - 1$, then f is a local homeomorphism and thus in view of Theorem 2.3.2 a homeomorphism. So we may assume that $l \leq (n - 2)$.

In view of Proposition 2.3.10 and Theorem 2.3.2, it suffices to prove that f is a local homeomorphism at every point of $x_0 \in |\Sigma_l|$ which by part 4 of Proposition 2.2.6 is equivalent to the homeomorphism property of the piecewise linear function $f'(x_0; .)$. Recall from Proposition 2.2.6 that a conical partition corresponding to $f'(x_0; .)$ is given by the localization $\Sigma'(x_0)$. Part 1 of Lemma 2.3.4 shows that the k-th branching number of $\Sigma'(x_0)$ does not exceed the k-th branching number of Σ. Hence the assertion of the theorem holds for piecewise affine functions if it holds for their B-derivatives which are piecewise linear functions. So we may assume without loss of generality that f is a piecewise linear function and that Σ is a corresponding conical subdivision of $\mathbb{R}^n$. The result is proved by induction over the dimension n.

If $n = 2$, then our assumption $l \leq (n-2)$ shows that Σ is pointed and that $k = 2$. Hence Σ consists of four pointed polyhedral cones. Since f is coherently oriented, every pointed cone is mapped by f onto a pointed cone. In view of Corollary 2.3.1, the statement for $n = 2$ is thus an immediate consequence of the elementary fact that it is not possible to cover $\mathbb{R}^2$ twice with four pointed cones.

Now suppose that $n \geq 3$ and that the claim holds for all piecewise linear functions $f : \mathbb{R}^q \to \mathbb{R}^q$ with $2 \leq q \leq (n - 1)$. We distinguish two cases:

1. Suppose first that $k = n$ and that the $(n - 1)$-st branching number is greater than $2(n - 1)$. In this case, Σ is pointed since by assumption $n = k \leq n - l$. Moreover, since the nth branching number does not exceed $2n$, the collection Σ consists of at most $2n$ cones. Since the $(n - 1)$-st branching number is greater than $2(n - 1)$, there exists an extremal ray in Σ_1 which is contained in at least $2n - 1$ cones, say $\sigma_1, \ldots, \sigma_{2n-1}$. Let v be a unit generator of this extremal ray. Since f is coherently oriented, $f(v) \neq 0$ and the images $f(\sigma)$ of the pointed cones $\sigma \in \Sigma$ are pointed as well. Hence the vector $-f(v)$ is not contained in any of the pointed cones $f(\sigma_i)$, $i = 1, \ldots, 2n - 1$ since by assumption any of the latter cones contains the vector $f(v)$. Since the finite union of closed sets is closed, there exists an open neighborhood U of $-f(v)$ which does not intersect the union of the cones $f(\sigma_i)$, $i = 1, \ldots, 2n - 1$. Since a coherently oriented piecewise linear mapping is surjective and Σ consists of not more than $2n$ cones, every vector $y \in U$ has at least one preimage and all preimages of the vectors $y \in U$ are contained in the remaining cone $\sigma_{2n} \in \Sigma$. Since f is coherently oriented, σ_{2n} contains at most one preimage of $y \in U$ and hence Corollary 2.3.1 shows that f is a homeomorphism.
2. If $k = n$ and the $(k - 1)$-st branching number does not exceed $2(k - 1)$, then we may replace k by $(k - 1)$ without violating the assumptions of the theorem.

It thus remains to prove the assertion for the case $k \leq (n-1)$. If, on the one hand, Σ is pointed, then Theorem 2.3.2 shows that it suffices to prove that $f'(x; .)$ is a homeomorphism at every point $x \in |\Sigma_1|\backslash|\Sigma_0|$. If, on the other hand, $l \geq 1$, then Proposition 2.3.10 and Theorem 2.3.2 show that it suffices to prove that $f'(x; .)$ is a homeomorphism at every point $x \in \Sigma_l$. In any case, Proposition 2.2.6 shows that the corresponding subdivisions $\Sigma'(x)$ have a lineality space L with dimension $1 \leq p \leq n - k$. Since f is coherently oriented, Lemma 2.3.2 shows that f maps L homeomorphically onto $f'(x; L)$. In view of Lemma 2.3.3 it thus suffices to show that the function $g(y) = \Pi_{f'(x;L)^\perp}(f'(x; y))$ maps $L^\perp$ homeomorphically onto $f'(x; L)^\perp$. Proposition 2.3.12 shows that g is coherently oriented with corresponding subdivision $\tilde{\Sigma}$ of $L^\perp$. Lemma 2.3.4 states that the subdivision $\tilde{\Sigma}$ inherits from Σ the property that the k-th branching number does not exceed $2k$. Applying nonsingular linear transformations, we may assume that $L^\perp = f'(x; L)^\perp = \mathbb{R}^{n-p}$. Hence the induction assumption shows that g maps $L^\perp$ homeomorphically onto $f'(x; L)^\perp$ and thus f is a local homeomorphism at x which in view of Theorem 2.3.2 completes the proof. $\qquad\square$

2.3.5 Comments and References

It has already been observed in the early papers dealing with piecewise affine mappings that the determinant signs of the matrices corresponding to the selection functions play an important role in connection with the homeomorphism problem (cf. Theorem 2.3.1). Rheinboldt and Vandergraft have proved in [59] that a piecewise affine mapping is surjective, provided that all determinants of the matrices corresponding to a set of selection functions have the same nonvanishing sign (cf. also [8, 48]), and Schramm has shown in [70] that such mappings are open. In [33] Kuhn and öwen introduced the term coherent orientation for piecewise affine mappings which admit a collection of selection functions, the matrices of which have the same nonvanishing determinant sign, and showed that injective piecewise affine mappings are coherently oriented. They have also indicated the geometric meaning of coherent orientation which we have used as a definition (cf. Proposition 2.3.5).

The problem of local versus global homeomorphisms has been intensively studied in the literature. The general theorems on covering maps and local homeomorphisms of Sect. 2.3.2 remain valid for arbitrary metric spaces X and Y. Theorem 2.3.3 is due to Browder (cf. [6, 33], Sect. 4.2). Further conditions which ensure that a local homeomorphism is a covering map are given in [50]. Theorems 2.3.4 and 2.3.5 can be found in Spanier's book [78] as indicated. The introduction of covering map theory to the study of piecewise affine functions is due to Schramm (cf. [70]) and has been further exploited in [33] to prove the branching number theorem for the important special case of the second branching number. The general branching number theorem has been proved in [72].

The factorization results of Sect. 2.3.3 can be found in the article [16] of Eaves and Rothblum. The latter paper extends most of the results which we have presented in this section to piecewise affine mappings over general ordered fields.

Sufficient homeomorphism conditions which rely on the matrix-vector pairs rather than the subdivision can be found e.g., in the articles [20] of Fujisawa and Kuh, [30] of Kojima and Saigal, and in the recent paper [58].

2.4 Euclidean Projections

To illustrate the results developed in the present chapter, we proceed with an analysis of Euclidean projections onto polyhedra. This class of mappings is a particularly important subclass of piecewise affine functions and has numerous applications, one of which is outlined at the end of this section.

The *Euclidean projection* Π_S onto a closed convex set $S \subseteq \mathbb{R}^n$ is the mapping which assigns to each point $x \in \mathbb{R}^n$ the unique point $\Pi_S(x) \in S$ which has minimal Euclidean distance to x. The existence of a minimizer follows from Weierstass' Theorem which implies that a continuous function which has compact lower level sets has a minimizer on every closed set. The uniqueness is a consequence of the convexity of S and the trivial fact that the midpoint between any two distinct points on a sphere has smaller distance to the center of the sphere than the endpoints. As an illustration of the mapping Π_S, we first calculate the Euclidean projection onto a line segment

$$[v, w] = \{tv + (1 - t)w | 0 \le t \le 1\}.$$

Since the result will be used in the sequel, it is convenient to formulate it as a lemma.

Lemma 2.4.1. *If v and w are distinct vectors in $\mathbb{R}^n$, then*

$$\Pi_{[v,w]}(x) = \begin{cases} v & \text{if } (x - v)^T (w - v) \le 0 \\ v + \dfrac{(x - v)^T (w - v)}{(w - v)^T (w - v)} (w - v) & \text{if } 0 < (x - v)^T (w - v) < (w - v)^T (w - v) \\ w & \text{if } (w - v)^T (w - v) \le (x - v)^T (w - v) \end{cases}$$

Proof. Elementary geometric arguments show that the unique point of the line aff$[v, w]$ which has minimal Euclidean distance to x is the point $v + \frac{(x-v)^T(w-v)}{(w-v)^T(w-v)}(w-v)$. Hence if $0 < \frac{(x-v)^T(w-v)}{(w-v)^T(w-v)} < 1$, then the latter point is the closest point to x on the line segment $[v, w]$. Clearly the endpoint v is the closest point to x on the line segment $[v, w]$ if the angle between $(x - v)$ and $(w - v)$ is obtuse, i.e., if $(x - v)^T (w - v) \le 0$, while w is the Euclidean distance minimizer to x if the angel between $(x - w)$ and $(v - w)$ is obtuse, i.e., if $(x - w)^T (v - w) \le 0$. The identity

$$(x - w)^T (v - w) = (w - v)^T (w - v) - (x - v)^T (w - v)$$

thus proves the assertion.   □

We note that for every closed convex set $S \subseteq \mathbb{R}^n$ the equation $\Pi_S(x) = s$ holds if and only if $\Pi_{[s,s']}(x) = s$ for every $s' \in S$. The next result establishes the nonexpansiveness of the Euclidean projection with respect to the Euclidean norm.

Proposition 2.4.1. *Let $S \subseteq \mathbb{R}^n$ be a nonempty closed convex set and $x, y \in \mathbb{R}^n$. Then*

$$\|\Pi_S(x) - \Pi_S(y)\| \le \|x - y\|.$$

Proof. Define $v = \Pi_S(x)$ and $w = \Pi_S(y)$. If $v = w$, then the inequality holds trivially. If $v \ne w$, then the convexity of S implies that $v = \Pi_{[v,w]}(x)$ and $w = \Pi_{[v,w]}(y)$ and hence Lemma 2.4.1 yields that

$$(x - v)^T (w - v) \le 0$$

and

$$(y - w)^T (v - w) \le 0.$$

Summing up both inequalities, we obtain

$$(v - w)^T (v - w) \le (x - y)^T (v - w).$$

Schwarz's inequality thus yields

$$\|v - w\|^2 \le \|x - y\|\|v - w\|$$

which establishes the assertion of the proposition.   □

Beside the definition of the Euclidean projection by means of a minimal principle there is an equivalent "dual" characterization of this mapping based on the normal cone concept.

Proposition 2.4.2. *For every nonempty closed convex set $S \subseteq \mathbb{R}^n$ and every $s \in S$ the following identity holds:*

$$\Pi_S^{-1}(s) = \{s\} + N_S(s).$$

Proof. We have to prove that $\Pi_S(x) = s$ if and only if $(x - s) \in N_S(s)$. By definition of the normal cone a vector v is an element of $N_S(s)$ if and only if the set S is a subset of the halfspace $H(s, v) = \{y \in \mathbb{R}^n | v^T y \le v^T s\}$. We thus have to prove that for every $x \in \mathbb{R}^n$ the equality $\Pi_S(x) = s$ holds if and only if $S \subseteq H(s, x - s)$. To see this, recall that $\Pi_S(x) = s$ if and only $\Pi_{[s,s']}(x) = s$ for every $s' \in S$. Lemma 2.4.1 shows that the latter statement holds if and only if $(x - s)^T s' \le (x - s)^T s$ for every $s' \in S$, i.e., if and only if $S \subseteq H(s, x - s)$.   □

2.4.1 The Euclidean Projection onto a Polyhedron

Next we show that the Euclidean projection onto a polyhedron is piecewise affine. Before we present this result, we calculate the Euclidean projection onto an affine subspace S of $\mathbb{R}^n$, hereby using the result of Proposition 2.4.2. Recall that any affine subspace S is representable as the intersection of a finite number of $(n-1)$-dimensional hyperplanes with linearly independent normals.

Proposition 2.4.3. *If $S = \{x \in \mathbb{R}^n | Ax = b\}$, where A is an $m \times n$-matrix with full row rank and $b \in \mathbb{R}^m$, then $\Pi_S(z) = (I - A^T(AA^T)^{-1}A)z + (A^T(AA^T)^{-1})b$.*

Proof. By Proposition 2.4.2 the vector $x \in S$ is the Euclidean projection of the vector $z \in \mathbb{R}^n$ if and only if

$$z \in \{x\} + N_S(x).$$

It is an immediate consequence of the definition that the normal cone of the affine subspace $S \subseteq \mathbb{R}^n$ at a point $x \in S$ is the orthogonal complement of the linear subspace $S - x$ which is independent of x. In view of the representation of S, we thus obtain

$$N_S(x) = \{A^T y | y \in \mathbb{R}^m\}.$$

Hence the Euclidean projection $\Pi_S(z)$ can be determined by solving the equation

$$\begin{pmatrix} z \\ b \end{pmatrix} = \begin{pmatrix} I & A^T \\ A & 0 \end{pmatrix} \begin{pmatrix} \Pi_S(z) \\ y \end{pmatrix},$$

and thus the assertion of the theorem follows from the identity

$$\begin{pmatrix} I & A^T \\ A & 0 \end{pmatrix}^{-1} = \begin{pmatrix} I - A^T(AA^T)^{-1}A & A^T(AA^T)^{-1} \\ (AA^T)^{-1}A & -(AA^T)^{-1} \end{pmatrix}.$$

$\square$

The latter proposition shows that the Euclidean projection onto an affine space is an affine function. In order to prove that the Euclidean projection onto a polyhedron P is a piecewise affine function we recall from Proposition 2.1.3 that the faces of a polyhedron $P = \{x \in \mathbb{R}^n | Ax \le b\}$ can be indexed by elements of the collection

$$\mathscr{I}(A,b) = \{I \subseteq \{1,\dots,m\} | \text{there exists a vector } x \in \mathbb{R}^n \text{ with}$$

$$a_i^T x = b_i, i \in I, a_j^T x < b_j, j \in \{1,\dots,m\}\setminus I\}, \tag{2.42}$$

where the vectors $a_1,\dots,a_m$ form the rows of the $m \times n$-matrix A, while the numbers $b_1,\dots,b_m$ are the components of the vector $b \in \mathbb{R}^m$. The collection of all sets

$$F_I = \{x \in \mathbb{R}^n | a_i^T x = 0, i \in I\}, I \in \mathscr{I}(A,b), \tag{2.43}$$

is the collection of all nonempty faces of P. Moreover, Proposition 2.1.2 and part 3 of Proposition 2.1.3 show that the normal cone of P at a relative interior point x of the face F_I is given by

$$N_I = \text{cone}\{a_i \,|\, i \in I\}, \tag{2.44}$$

where $N_\emptyset = \{0\}$ by definition.

Proposition 2.4.4. *Let* $P = \{x \in \mathbb{R}^n \,|\, Ax \leq b\}$, *where A is an $m \times n$-matrix and $b \in \mathbb{R}^m$, let $\mathscr{I}(A, b)$, F_I, and N_I be defined by (2.42)–(2.44), respectively. If $z \in F_I + N_I$ for some $I \in \mathscr{I}(A, b)$, then $\Pi_P(z) = \Pi_{S_I}(z)$, where $S_I = \{x \in \mathbb{R}^n \,|\, a_i^T x = b_i, i \in I\}$.*

Proof. If $z \in F_I + N_I$, then $z \in \{x\} + N_I$ for some $x \in F_I$. Since, on the one hand, Proposition 2.1.2 implies that $N_I \subseteq N_P(x)$, we obtain $z \in \{x\} + N_P(x)$ and, in view of Proposition 2.4.2, $x = \Pi_P(z)$. Since, on the other hand, $N_I \subseteq N_{S_I}(x) = \text{lin}\{a_i \,|\, i \in I\}$, we obtain $z \in \{x\} + N_{S_I}(x)$ and since $x \in S_I$, Proposition 2.4.2 shows that $x = \Pi_{S_I}(z)$. Thus $\Pi_P(z) = \Pi_{S_I}(z)$. $\square$

2.4.2 The Normal Manifold

The *normal manifold* induced by the polyhedron $P = \{x \in \mathbb{R}^n \,|\, Ax \leq b\}$ is defined to be the collection of all polyhedra $P_I = F_I + N_I$, $I \in \mathscr{I}(A, b)$, where $\mathscr{I}(A, b)$, F_I, and N_I are defined by (2.42)–(2.44), respectively. Note that the normal manifold is independent of the representation of the polyhedron since every set P_I is the sum of a face F_I of P and the normal cone N_I of P at a relative interior point of this face. We will see below that the normal manifold is a polyhedral subdivision and that its second branching number is 4. To prove this result, we use the following lemma.

Lemma 2.4.2. *1. If $I \in \mathscr{I}(A, b)$, then every face of the polyhedron F_I is of the from F_J, for an index set $J \in \mathscr{I}(A, b)$ with $I \subseteq J$.*

2. If $I \in \mathscr{I}(A, b)$, then every proper face of the cone N_I is of the form N_J, $J \in \mathscr{I}(A, b)$, $J \subseteq I$, $J \neq I$.

3. If $I, J \in \mathscr{I}(A, b)$ are two distinct index sets such that $P_I \cap P_J \neq \emptyset$, then

(a) $P_I \cap P_J = F_I \cap F_J + N_I \cap N_J$,

(b) $F_I \cap F_J$ is a common proper face of F_I and F_J,

(c) $N_I \cap N_J = N_{I \cap J}$ is a common proper face of N_I and N_J.

Proof. 1. The first assertion is a trivial consequence of the fact that every face of F_I is representable as $\{x \in F_I \,|\, a_k^T x = 0, k \in K\}$ for some $K \subseteq \{1, \dots, m\} \setminus I$. Taking the maximal index sets K, we thus conclude that every face of F_I is indeed of the form F_J, $J \subseteq I$, $J \in \mathscr{I}(A, b)$.

2. To see the second assertion, note that every proper face F of N_I is generated by a subset of the vectors a_i, $i \in I$. So let $J \subseteq I$ be a maximal subset of I such that $F = N_J$. Since F is proper, $J \neq I$. In order to see that $J \in \mathscr{I}(A, b)$, recall that every face is a max-face, i.e., there exists a vector v such that the points of F are the maximizers of $v^T x$ over $x \in N_I$. Since J was assumed to be maximal, we conclude that $v^T a_j = 0$ for every $j \in J$, and $v^T a_k < 0$ for every $k \in \{1, \ldots, m\} \backslash J$, which shows that $J \in \mathscr{I}(A, b)$. To see the converse let $J \in \mathscr{I}(A, b)$ with $J \subseteq I$, $J \neq I$. In particular, there exists a vector $v \in \mathbb{R}^n$ such that $v^T a_j = 0$ for every $j \in J$ and $v^T a_k < 0$ for every $k \in \{1, \ldots, m\} \backslash J$. Hence the set N_J is the set of maximizers of the linear function $v^T x$ over $x \in N_I$ and thus a face of N_I. The face is proper since for every $i \in I \backslash J$ the strict inequality $v^T a_i < 0$ holds.

3. (a) Note first that by the definition of the sets P_I and P_J the inclusion

$$(F_I \cap F_J) + (N_I \cap N_J) \subseteq P_I \cap P_J$$

holds. To prove the converse inclusion, we note that $z \in P_I \cap P_J$ if and only if there exist two vectors $x \in F_I$ and $y \in F_J$ such that

$$z \in (\{x\} + N_I) \cap (\{y\} + N_J). \tag{2.45}$$

Since by Proposition 2.1.2 the inclusion $N_I \subseteq N_P(x)$ holds for every $x \in F_I$, (2.45) implies

$$z \in (\{x\} + N_P(x)) \cap (\{y\} + N_P(y)), \tag{2.46}$$

which in view of Proposition 2.4.2 is equivalent to $x = \Pi_P(z) = y$ and thus proves that

$$\Pi_P(z) \in F_I \cap F_J. \tag{2.47}$$

Moreover, since $x = \Pi_P(z) = y$, (2.45) also shows that

$$z - \Pi(z) \in N_I \cap N_J,$$

which proves the inclusion $P_I \cap P_J \subseteq (F_I \cap F_J) + (N_I \cap N_J)$.

(b) Recall that (2.47) holds for every $z \in P_I \cap P_J$. Since the latter set was assumed to be nonempty, we thus conclude that $F_I \cap F_J \neq \emptyset$. The assertion thus follows from the fact that the nonempty intersection of any two distinct faces F_I and F_J of P is a proper face of both faces F_I and F_J.

(c) Note first that the inclusion $N_{I \cap J} \subseteq N_I \cap N_J$ is trivial. To see the converse inclusion, let $v, w \in \mathbb{R}^n$ be two vectors satisfying

$$a_i^T v = b_i, i \in I, a_k^T v < b_k, k \in \{1, \dots, m\} \backslash I,$$

$$a_j^T w = b_j, j \in J, a_l^T v < b_l, l \in \{1, \dots, m\} \backslash J. \tag{2.48}$$

Note that $y^T v = 0$ for every $y \in N_I$ and $y^T w = 0$ for every $y \in N_J$ and thus

$$y^T (v + w) = 0 \quad \text{for every } y \in N_I \cap N_J. \tag{2.49}$$

If $y \in N_I$, then $y = \sum_{i \in I \cap J} \lambda_i a_i + \sum_{j \in I \backslash J} \lambda_j a_j$, for some multipliers $\lambda_i \geq 0$. If $y \notin N_{I \cap J}$, then at least one of the multipliers λ_j, $j \in I \backslash J$, is nonzero and hence $y^T (v + w) < 0$. In view of (2.49), we deduce that $y \notin N_I \cap N_J$ and thus $N_I \cap N_J \subseteq N_{I \backslash J}$. Since the same holds with the roles of I and J interchanged, we conclude that $N_I \cap N_J \subseteq N_{I \cap J}$. The assertion that $N_{I \cap J}$ is a common proper face of N_I and N_J is an immediate consequence of part 2 and the trivial observation that $I \cap J \in \mathscr{I}(A, b)$ for every $I, J \in \mathscr{I}(A, b)$. $\qquad \square$

With the aid of the latter lemma, one easily proves the following important result.

Proposition 2.4.5. *The normal manifold induced by a polyhedron P is a polyhedral subdivision of $\mathbb{R}^n$, the second branching number of which is 4.*

Proof. Let $P = \{x \in \mathbb{R}^n | Ax \leq b\}$. The full dimensionality of the sets $F_I + N_I$, $I \in \mathscr{I}(A, b)$ is an immediate consequence of the fact that F_I contains a relative interior point x and that the set N_I spans the subspace $\mathrm{lin}\{a_i | i \in I\}$, while the set $F_I - \{x\}$ spans its orthogonal complement. The covering property is also easily verified since for every $z \in \mathbb{R}^n$ there exists a unique face F_I of P containing the point $x = \Pi_P(z)$ in its relative interior. Since by Proposition 2.4.2 the identity $x = \Pi_P(z)$ holds if and only if $z \in \{x\} + N_P(x)$ and by Proposition 2.1.2 $N_P(x) = N_I$ for every relative interior point x of F_I, we conclude that $z \in F_I + N_I$. The fact that the nonempty intersection of any two polyhedra P_I and P_J of the normal manifold is a common face of both polyhedra is a direct consequence of Lemma 2.1.2 and part 3 of Lemma 2.4.2. We thus conclude that the normal manifold induced by a polyhedron is indeed a polyhedral subdivision.

To see that the second branching number is 4, note that Lemmas 2.1.2 and 2.4.2 show that every face of a polyhedron $P_I = F_I + N_I$, $I \in \mathscr{I}(A, b)$, is of the form $F = F_J + N_K$, where $J, K \in \mathscr{I}(A, b)$ with $K \subseteq I \subseteq J$, and, moreover, that $F_J + N_K \subseteq P_{\tilde{I}}$ for some index set $\tilde{I} \in \mathscr{I}(A, b)$ if and only if $K \subseteq \tilde{I} \subseteq J$. Since $K \subseteq J$, the sets F_J and N_K are contained in mutually orthogonal subspaces of $\mathbb{R}^n$ and thus $\dim(F_J + N_K) = \dim F_J + \dim N_K$. Since $K \in \mathscr{I}(A, b)$, we deduce from the definition of N_K and F_K that $\dim N_K = (n - \dim F_K)$. Hence $\dim(F_J + N_K) = (n - 2)$ if and only if $\dim F_J = (\dim F_K - 2.)$ Since $K \subseteq I \subseteq J$ for $I \in \mathscr{I}(A, b)$ if and only if F_J and F_I are faces of the polyhedron F_K and F_J is a face of F_I, the proof is complete if we can show that every face of dimension $(\dim F_K - 2)$ of the polyhedron F_K is contained in precisely two faces of dimension $(\dim F_K - 1)$.

To see this, recall that

$$F_K = \left\{ x \in \mathbb{R}^n | a_k^T x = b_k, k \in K, a_j^T x \leq b_j, j \in \{1,\ldots,m\}\backslash K \right\}.$$

Since $K \in \mathscr{I}(A,b)$, there exists a vector $v \in \mathbb{R}^n$ with

$$a_k^T v = b_k, k \in K, a_j^T v < b_j, j \in \{1,\ldots,m\}\backslash K.$$

To simplify the exposition, we make some additional assumptions which do not affect the face lattice of F_K. First of all, since the face lattice of a translation of a polyhedron is the collection of all translated faces, we may assume without loss of generality that $v = 0$ and thus

$$\begin{aligned} b_k &= 0 \quad \text{for every } k \in K \\ b_j &> 0 \quad \text{for every } j \in \{1,\ldots,m\}\backslash K. \end{aligned} \tag{2.50}$$

Secondly, since any vector in $\mathbb{R}^n$ can be uniquely decomposed as the sum of a vector in a linear subspace and of a vector in its orthogonal complement, any vector a_j, $j \notin K$, is the sum of a vector $c_j \in \text{lin}\{a_k | k \in K\}$ and a vector $d_j \in \{x \in \mathbb{R}^n | a_k^T x = 0, k \in K\}$. If $j \notin K$ is an arbitrary index and $x \in \mathbb{R}^n$ is a vector with $a_k^T x = b_k$ for every $k \in K$, then, in view of (2.50), the vector x satisfies the inequality $a_j^T x \leq b_j$ if and only if it satisfies the inequality $d_j^T x \leq b_j$. Since the faces of the polyhedron F_K do not depend on its representation, we may replace a_j by d_j, if necessary, and thus assume without loss of generality that

$$a_j \in \text{lin}\{a_k | k \in K\}^\perp \quad \text{for every } j \in \{1,\ldots,m\}\backslash K. \tag{2.51}$$

Finally, we may delete redundant inequalities from the representation of F_K and assume that for each $j \in \{1,\ldots,m\}$ the constraint $a_j^T x \leq b_j$ is necessary, i.e., the set of feasible solutions increases if the inequality is removed. Note that every face of dimension (dim $F_K - 1$) is of the form $F_{K\cup\{j\}}$, where $j \notin K$. In order to prove the assertion that every face of dimension (dim $F_K - 2$) is contained in precisely two faces of dimension (dim $F_K - 1$), it thus suffices to show that the dimension of the nonempty intersection of any three distinct faces of the from $F_{K\cup\{j\}}$, $j \notin K$, is less than (dim $F_K - 2$). In view of (2.51), it suffices to check that any three vectors a_{j_1}, a_{j_2}, and a_{j_3} with $j_i \in \{1,\ldots,m\}\backslash K$ are linearly independent, provided there exists a vector $x \in F_K$ with

$$a_{j_i}^T x = b_{j_i}, i = 1,2,3. \tag{2.52}$$

If two vectors are linearly dependent, say $a_{j_1} = \alpha a_{j_2}$, then we deduce from (2.50) and (2.52) that $\alpha > 0$. Hence one of the constraints is redundant, which contradicts the assumption that all inequalities are necessary. Therefore any two vectors are linearly independent and the equation

$$\lambda_1 a_{j_1} + \lambda_2 a_{j_2} + \lambda_3 a_{j_3} = 0$$

implies that either all multipliers λ_i vanish or all of them are nonzero. If one of the multipliers, say λ_1, is positive, while another multiplier, say λ_2, is negative, then either $a_{j_1} \in \mathrm{cone}\{a_{j_2}, a_{j_3}\}$ or $a_{j_2} \in \mathrm{cone}\{a_{j_1}, a_{j_3}\}$, depending on the sign of λ_3. Both implications contradict the non-redundancy assumption in view of (2.52). Therefore the multipliers are either all positive or all negative. However, if this is the case, then $a_{j_1} \in \mathrm{cone}\{-a_{j_2}, -a_{j_3}\}$, which in view of (2.52) contradicts the assumption (2.50) and thus proves the linear independence of the vectors a_{j_1}, a_{j_2}, and a_{j_3}.   □

We note that we have actually shown in the final part of the latter proof that every ridge of a polyhedron is contained in exactly two facets, a fact which is well known in polyhedral theory. This is the geometric reason why the second branching number of the normal manifold induced by a polyhedron is 4.

2.4.3 An Application: Affine Variational Inequalities

To illustrate the results developed in the present section, we show how they can be used to answer existence and uniqueness questions for the solution of an affine variational inequality

$$(Cx + d)^T x \ge (Cx + d)^T y \quad \text{for every } y \in P = \{z \in \mathbb{R}^n \,|\, Az \le b\}, \qquad (2.53)$$

where A and C are $m \times n$- and $n \times n$-matrices, respectively, $b \in \mathbb{R}^m$, and $d \in \mathbb{R}^n$. In Sect. 1.1.1 we have indicated how variational inequalities can be used to model equilibrium situations of dynamical systems.

Let us first show how the variational inequality (2.53) can be turned into a piecewise affine equation. Note that the definition (2.2) of the normal cone mapping N_P shows that x solves the variational inequality (2.53) if and only if $Cx + d \in N_P(x)$ which in view of Proposition 2.4.2 is equivalent to the equation

$$\Pi_P(x + Cx + d) = x. \qquad (2.54)$$

Using the results developed in the former sections, one easily proves the following existence and uniqueness result.

Theorem 2.4.1. *Let A and C be $m \times n$- and $n \times n$-matrices, respectively, $b \in \mathbb{R}^m$, $d \in \mathbb{R}^n$, let $a_1, \dots, a_m$ be the row vectors of A, and let $\mathscr{I}(A, b)$ be defined by (2.6). Choose for every $I \in \mathscr{I}(A, b)$ a maximal set of linearly independent vectors from $\{a_i \,|\, i \in I\}$ and store these vectors in the rows of a matrix M_I. If all matrices*

$$\left(I - M_I^T (M_I M_I^T)^{-1} M_I\right) C - M_I^T (M_I M_I^T)^{-1} M_I, \ i \in \mathscr{I}(A, b),$$

have the same nonvanishing determinant sign, then the variational inequality (2.53) has a unique solution.

Proof. Since the variational inequality (2.53) is equivalent to the (2.54), it suffices to prove that the function $\Pi_P(x + Cx + d) - x$ has a unique zero. In view of Propositions 2.4.4 and 2.4.5, the mapping $\Pi_P(x + Cx + d) - x$ is piecewise affine with affine selection functions

$$\Pi_{S_I}(x + Cx + d) - x.$$

Moreover, the normal manifold induced by P is a polyhedral subdivision corresponding to the mapping $\Pi_P(x + Cx + d) - x$. In view of the assumptions and Proposition 2.4.3, we deduce that $\Pi_P(x + Cx + d) - x$ is coherently oriented, and in view of Theorem 2.3.7 and Proposition 2.4.5, the latter mapping thus has a unique zero.   □

Specifying the polyhedron P, we can use the latter theorem to develop the well-known existence and uniqueness conditions for the *linear complementarity problem* of finding a vector $x \in \mathbb{R}^n$ which satisfies

$$x \geq 0, \quad Sx + r \geq 0, \quad (Sx + r)^T x = 0, \tag{2.55}$$

where S is an $n \times n$-matrix and $r \in \mathbb{R}^n$. Note that (2.55) is equivalent to

$$\max\{-x, -Sx - r\} = 0.$$

Since $\Pi_{\mathbb{R}^n_+}(z) = \max\{z, 0\}$, the complementarity problem is equivalent to (2.54) with $C = -S$, $d = -r$, and $P = \mathbb{R}^n_+$, i.e., $A = \mathrm{I}$ and $b = 0$ in the formulation (2.53). Since $\mathscr{I}(\mathrm{I}, 0)$ is the collection of all subsets of $I \subseteq \{1, \ldots, n\}$, the matrices M_I in the above theorem contain the row vectors e_i, $i \in I$, where e_i is the ith unit vector in $\mathbb{R}^n$. Hence the matrix $D^I = M_I^T(M_I M_I^T)^{-1}M_I$ is a diagonal matrix with diagonal entries

$$D_{ii}^I = \begin{cases} 1 & \text{if } i \in I, \\ 0 & \text{otherwise.} \end{cases}$$

The rows N_i^I of the matrix

$$N^I = \left(\mathrm{I} - M_I^T(M_I M_I^T)^{-1}M_I\right)C - M_I^T(M_I M_I^T)^{-1}M_I$$

are therefore given by

$$N_i^I = \begin{cases} -e_i & \text{if } i \in I, \\ -S_i & \text{otherwise,} \end{cases} \tag{2.56}$$

where S_i denotes the ith row of the matrix $S = -C$. Since $N^I = -\mathrm{I}$ for $I = \{1, \ldots, n\}$, the assumption of the above theorem is satisfied if all determinants of the matrices N^I, $I \subseteq \{1, \ldots, n\}$, are negative. In view of the Laplace expansion and the special form (2.56) of the matrices N^I, this is equivalent to the requirement that all principal minors of the matrix S are positive. In the terminology of linear complementarity theory, a matrix with the latter property is called a *P-matrix*. Summing up, we have proven the following well-known result as a corollary of Theorem 2.4.1:

Corollary 2.4.1. *If S is a P-matrix, then the linear complementarity problem (2.55) has a unique solution.*

2.4.4   Comments and References

For a detailed account on the Euclidean projection, we refer to Zarantonello's article [81] and to Singer's book [77]. By a result of Pang the Euclidean projection Π_P can be locally decomposed as

$$\Pi_P(x + y) = \Pi_P(x) + \Pi_{C(x)}(y),$$

where $C(x) = \text{cone}\{z \in P - \{\Pi_P(x)\} | (\Pi_P(x) - x)^T x = 0\}$ is the so-called critical cone at $\Pi_P(x)$ (cf. Lemma 4 in [53] and Corollary 4.5 in [61]). In particular, the latter result implies that the B-derivative of Π_P is given by

$$\Pi_P'(x; y) = \Pi_{C(x)}(y).$$

The Euclidean projection onto polyhedra has been further studied by Robinson in [66], where the normal manifold is introduced and analyzed in connection with normal maps induced by linear transformations (cf. also [55]). The branching number theorem has been first proved by Ralph in [56]. His proof differs from ours by the use of purely geometric arguments. In particular, he does not exploit the representation of the polyhedron as a system of linear inequalities (cf. [71]).

The results of Theorem 2.4.1 are related to the results of Robinson in [66], where the affine variational inequality is transformed into a normal map. For a first account on the well-known existence and uniqueness result for the linear complementarity problem (cf. Corollary 2.4.1), we refer to the paper [69] of Samelson, Thrall, and Wesler. A comprehensive survey of the linear complementarity problem is given in Murty's book [47].

2.5   Appendix: The Recession Function

In this appendix to the chapter on piecewise affine functions, we introduce the recession function and study the relationship between a piecewise affine function and its *recession function*. The recession function of a function $f : \mathbb{R}^n \to \mathbb{R}^m$ is defined by

$$f^\infty(x) = \lim_{\alpha \to \infty} \frac{f(\alpha x)}{\alpha},$$

provided the limit exists for every $x \in \mathbb{R}^n$. If f is piecewise affine, then the latter limit always exist since for every x there exists a polyhedron σ in the corresponding polyhedral subdivision such that $\alpha x \in \sigma$ for every sufficiently large $\alpha > 0$. Hence if (A^i, b^i) is the corresponding matrix-vector pair, we obtain

$$f^\infty(x) = \lim_{\alpha \to \infty} \frac{A^i \alpha x + b^i}{\alpha} = A^i x.$$

In particular, the recession function of an affine function $f(x) = Ax + b$ is the linear function $f^\infty(x) = Ax$.

The recession function has some interesting properties which are reminiscent of the properties of the B-derivative. In particular, the recession function is piecewise linear. This is easily seen by considering a max–min representation of a real-valued piecewise affine function $f : \mathbb{R}^n \to \mathbb{R}$ (cf. Proposition 2.2.2), i.e.,

$$f(x) = \max_{1 \le i \le l} \min_{j \in M_i} a_j^T x + b_j,$$

where $a_1^T x + b_1, \ldots, a_k^T + b_k$ are selection functions corresponding to f and $M_i \subseteq \{1, \ldots, k\}, i = 1, \ldots, l$. The recession function of f is easily calculated as

$$f^\infty(x) = \max_{1 \le i \le k} \min_{j \in M_i} a_j^T x,$$

which is a piecewise linear function. Since a function $f : \mathbb{R}^n \to \mathbb{R}^m$ is piecewise linear if and only if its component functions are piecewise linear, we obtain the following result.

Proposition 2.5.1. *The recession function of a piecewise affine function is piecewise linear.*

Beside its piecewise linearity, the recession function has a second property in common with the B-derivative, namely the existence of a chain-rule.

Proposition 2.5.2. *If $f : \mathbb{R}^n \to \mathbb{R}^m$ and $g : \mathbb{R}^k \to \mathbb{R}^n$ are piecewise affine functions, then*

$$(f \circ g)^\infty = f^\infty \circ g^\infty.$$

Proof. Let us fix some vector $x \in \mathbb{R}^k$ and suppose

$$g(\alpha x) = A^i \alpha x + b^i$$

for every sufficiently large $\alpha > 0$, i.e., $g^\infty(x) = A^i x$. Note that the following two limits both exist:

$$(f \circ g)^\infty(x) = \lim_{\alpha \to \infty} \frac{f(A^i \alpha x + b^i)}{\alpha},$$

$$(f^\infty \circ g^\infty)(x) = \lim_{\alpha \to \infty} \frac{f(\alpha A^i x)}{\alpha}.$$

Since f is Lipschitzian (cf. Proposition 2.2.7), there exists a constant L such that

$$\left\| \frac{f(A^i \alpha x + b^i)}{\alpha} - \frac{f(A^i \alpha x)}{\alpha} \right\| \leq L \left\| \frac{b^i}{\alpha} \right\|.$$

Letting α tend to infinity, we conclude that $(f \circ g)^\infty(x) = (f^\infty \circ g^\infty)(x)$. □

The main result of this appendix states that the recession function f^∞ of a coherently oriented piecewise affine function f preserves enough information to decide whether or not f is a homeomorphism.

Theorem 2.5.1. *A coherently oriented piecewise affine function is a homeomorphism if and only if its recession function is a homeomorphism.*

Proof. Let Σ be a polyhedral subdivision corresponding to the piecewise affine function $f : \mathbb{R}^n \to \mathbb{R}^n$. Recall that the recession cone of a polyhedron $\sigma \in \Sigma$ is the set of all vectors $y \in \mathbb{R}^n$ such that $\sigma + \mathrm{cone}\{y\} \in \sigma \subseteq \sigma$. Setting $g(x) = x - x_0$ for some vector $x_0 \in \sigma$, one readily deduces from the latter chain rule that $f^\infty(y) = Ay$, provided the function f coincides on σ with an affine function $Ay + b$. Since for every $y \in \mathbb{R}^n$ there exists a polyhedron $\sigma \in \Sigma$ such that $\alpha y \in \sigma$ for every sufficiently large $\alpha \geq 0$, every $y \in \mathbb{R}^n$ is contained in the recession cone of some polyhedron $\sigma \in \Sigma$. Hence if f is a piecewise affine function with corresponding collection of matrix-vector pairs $(A^1, b^1), \ldots, (A^k, b^k)$, then the collection of matrices $A^1, \ldots, A^k$ corresponds to the recession function f^∞. Moreover, if c is the recession cone of $\sigma \in \Sigma$, then $\sigma = \hat{\sigma} + c$ for some polytope $\hat{\sigma}$ and $f(\sigma) = f(\hat{\sigma}) + f^\infty(c)$, i.e., $f^\infty(c)$ is the recession cone of $f(\sigma)$. Let C_Σ be the collection of all n-dimensional recession cones of polyhedra $\sigma \in \Sigma$. We have already argued above that the collection of all recession cones of polyhedra in Σ covers $\mathbb{R}^n$. If we remove the cones with nonempty interior, this covering property is not affected; whence the cones in C_Σ cover $\mathbb{R}^n$. We also note that the intersection of the interiors of any two distinct cones c and $\tilde{c}$ is empty for otherwise the corresponding polyhedra $\sigma, \tilde{\sigma} \in \Sigma$ would have a common interior point. To see this, let $\sigma = \{x \in \mathbb{R}^n | Ax \leq b\}$ and $\tilde{\sigma} = \{x \in \mathbb{R}^n | \tilde{A}x \leq \tilde{b}\}$, where none of the rows of the matrices A and $\tilde{A}$ vanishes. In this case the intersection of the interiors of the corresponding recession cones c and $\tilde{c}$ is given by

$$\mathrm{int}(c) \cap \mathrm{int}(\tilde{c}) = \{x \in \mathbb{R}^n | Ax < 0, \tilde{A}x < 0\},$$

and one readily verifies that $\alpha x \in \mathrm{int}(\sigma) \cap \mathrm{int}(\tilde{\sigma})$ for every $x \in \mathrm{int}(c) \cap \mathrm{int}(\tilde{c})$ and every sufficiently large $\alpha > 0$.

Suppose first that f is not a homeomorphism and let $\tilde{\sigma}$ be an element of Σ with n-dimensional recession cone $\tilde{c}$. By Corollary 2.3.1

$$f(\tilde{\sigma}) \subseteq \bigcup_{\sigma \in \Sigma \setminus \{\tilde{\sigma}\}} f(\sigma). \tag{2.57}$$

If $x + \alpha y \in \tilde{\sigma}$ for every $\alpha \geq 0$, then the latter inclusion shows that there exists a polyhedron $\sigma \in \Sigma \setminus \{\tilde{\sigma}\}$ such that $x + \alpha y \in \sigma$ for every sufficiently large $\alpha > 0$. Hence y is contained in the recession cone of σ and thus the recession cone of the polyhedron $f(\tilde{\sigma})$ is a subset of the union of the recession cones $f(\sigma)$, $\sigma \in \Sigma \setminus \{\tilde{\sigma}\}$. Since $f^\infty(c)$ is the recession cone of $f(\sigma)$, we thus deduce from (2.57) that

$$f^\infty(\tilde{c}) \subseteq \bigcup_{c \neq \tilde{c}} f^\infty(c). \tag{2.58}$$

Since $\tilde{c}$ has nonempty interior and f^∞ is coherently oriented and coincides with a linear mapping on $\tilde{c}$, the cone $f^\infty(\tilde{c})$ has nonempty interior as well. Hence (2.58) implies that there exists at least one cone $c \in C_\Sigma \setminus \{\tilde{c}\}$ such that

$$\operatorname{int} f^\infty(c) \cap \operatorname{int} f^\infty(\tilde{c}) \neq \emptyset.$$

If v is an element of the latter set, then v has a preimage in the interior of c and in the interior of $\tilde{c}$. We have already seen above that the interiors of c and $\tilde{c}$ have no common points. Hence v has at least two preimages and f^∞ is not injective. Thus f is a homeomorphism if f^∞ is a homeomorphism.

The converse is proved in a similar way. If f is a homeomorphism, then

$$\operatorname{int} f(\sigma) \cap \operatorname{int} f(\tilde{\sigma}) = \emptyset$$

whenever $\sigma \neq \tilde{\sigma}$. Since the interiors of two polyhedra with n-dimensional recession cones have a nonempty intersection whenever the interiors of the recession cones have a nonempty intersection, we deduce that

$$\operatorname{int} f^\infty(c) \cap \operatorname{int} f^\infty(\tilde{c}) = \emptyset$$

for every two distinct recession cones $c, \tilde{c} \in C_\Sigma$. Since the union of all cones in C_Σ covers $\mathbb{R}^n$, we may use Corollary 2.3.1 to deduce that f^∞ is indeed a homeomorphism. $\square$

As an example of the simplification one can achieve in passing from a piecewise affine function to its recession function, we calculate the recession function of the Euclidean projection Π_P onto a convex polyhedron $P \subseteq \mathbb{R}^n$.

Proposition 2.5.3. *If P is a convex polyhedron with recession cone C, then* $\Pi_P^\infty = \Pi_C$.

Proof. Let P be decomposed as $P = Q + C$, where Q is a compact polyhedron and let $x \in \mathbb{R}^n$. Since Π_p is piecewise affine, the set $\Pi_P(\text{cone}\{x\})$ polygonal line, i.e., pieced together by a finite number of line segments. Hence there exists a direction $v \in \mathbb{R}^n$ and a positive real number $\hat{\alpha}$ such that

$$\Pi_P(\alpha x) = \Pi_P(\hat{\alpha}x) + (\alpha - \hat{\alpha})v \qquad (2.59)$$

for every $\alpha \geq \hat{\alpha}$. It follows from the definition of the recession function that $\Pi_P^\infty(x) = v$. Moreover, since $\Pi_P(\alpha x), \Pi_P(\hat{\alpha}x) \in P$, we conclude that $v \in C$. The definition of the Euclidean projection Π_P yields

$$\|\alpha x - \Pi_P(\alpha x)\| \leq \|\alpha x - y\|, \qquad (2.60)$$

for every $\alpha \geq \hat{\alpha}$ and every $y \in P$. Since P is the sum of Q and C, we may set $y = q + c$, $q \in Q$, $c \in C$, and since $\alpha C = C$ for every $\alpha > 0$, we obtain from (2.59) and (2.60) the inequality

$$\|\alpha x - \Pi_P(\hat{\alpha}x) + \hat{\alpha}v - \alpha v\| \leq \|\alpha x - q - \alpha c\| \qquad (2.61)$$

for every $q \in Q, c \in C$, and $\alpha \geq \hat{\alpha}$. Dividing both sides by α and letting α tend to infinity we conclude that $\|x - v\| \leq \|x - c\|$ for every $c \in C$, hence $\Pi_C(x) = v$.

$\square$

Chapter 3
Elements from Nonsmooth Analysis

3.1 The Bouligand Derivative

The success of the derivative concept for smooth functions relies on the following properties of the derivative:

- The function $g(x) = f(x_0) + \nabla f(x_0)(x - x_0)$ is a *first order approximation* of the function f at x_0, i.e. $g(x_0) = f(x_0)$ and

$$\lim_{x \to x_0} \frac{\| f(x) - g(x) \|}{\| x - x_0 \|} = 0.$$

- The first-order approximation, being an affine function, is considerably simpler than the original function.
- The existence of calculus rules allows the calculation of the derivative even for complicated composed functions.

In an attempt to generalize the derivative to a class of nonsmooth function, one has to try to preserve the latter properties in some sense. A natural extension of the classical derivative is the Bouligand derivative, a concept which we have already used in the latter chapter for a local analysis of piecewise affine functions.

Let $U \subseteq \mathbb{R}^n$ be an open set and let $f : U \to \mathbb{R}^m$ be a function. If for every $y \in \mathbb{R}^n$ the limit

$$f'(x_0; y) = \lim_{\substack{\alpha \to 0 \\ \alpha > 0}} \frac{1}{\alpha} (f(x_0 + \alpha y) - f(x_0))$$

exists, then f is called *directionally differentiable* at x_0 and the function $f'(x_0; .)$ is called the directional derivative of f at x_0. If f is directionally differentiable at x_0 and the function $g(x) = f(x_0) + f'(x_0; x - x_0)$ is a first-order approximation of f at x_0, then f is called *Bouligand differentiable* (B-differentiable) at x_0 and the function $f'(x_0; .)$ is called the B-derivative of f at x_0. The classical definition

S. Scholtes, *Introduction to Piecewise Differentiable Equations*, SpringerBriefs in Optimization, DOI 10.1007/978-1-4614-4340-7_3, © Stefan Scholtes 2012

of *Fréchet differentiability* (F-differentiability) is recovered by requiring that f is B-differentiable at x_0 and that $f'(x_0; .)$ is a linear function. In this case the function $f'(x_0; .)$ is called the F-derivative of f at x_0. We call the function $f : U \to \mathbb{R}^m$ B-differentiable (F-differentiable) if it is B-differentiable (F-differentiable) at all points of the open set U.

While the B-derivative concept preserves the first-order approximation property of the classical F-derivative, the approximation function is not necessarily affine any more. Nevertheless, it is still considerably simpler than the original function since the directional derivative of a function is positively homogeneous, i.e.,

$$f'(x_0; \lambda y) = \lambda f'(x_0; y) \text{ for every } \lambda \geq 0. \tag{3.1}$$

In fact, if $\beta = \alpha \lambda$ for a positive real number λ, then β tends to zero from above if and only if α does so; hence

$$\lim_{\substack{\alpha \to 0 \\ \alpha > 0}} \frac{1}{\alpha}(f(x_0 + \alpha \lambda y) - f(x_0)) = \lambda \lim_{\substack{\beta \to 0 \\ \beta > 0}} \frac{1}{\beta}(f(x_0 + \beta y) - f(x_0)).$$

The directional differentiability of a function is only of limited interest since the directional derivative provides a rather poor local approximation of the function. In fact, the function $f : \mathbb{R}^2 \to \mathbb{R}$ defined by

$$f(x_1, x_2) = \begin{cases} 0 \text{ if } x_2 \leq \sqrt{|x_1|}, \text{ or if } x_1 = 0, \\ 1 \text{ otherwise} \end{cases}$$

has a vanishing directional derivative at the origin without even being continuous at this point. The approximation properties of the B-derivative are much better. The proofs of the following facts are straightforward and therefore omitted.

Remark 3.1.1. 1. If $f : U \to \mathbb{R}^m$ is B-differentiable at $x_0 \in U$, then $f'(x_0; .)$ is continuous if and only if f is continuous.
2. If $f : U \to \mathbb{R}^m$ is a function, $x_0 \in U$, and $g : \mathbb{R}^n \to \mathbb{R}^m$ is a positively homogeneous function with

$$\lim_{y \to 0} \frac{\|f(x_0 + y) - f(x_0) - g(y)\|}{\|y\|} = 0,$$

then f is B-differentiable at x_0 and $f'(x_0; y) = g(y)$.
3. A function $f : U \to \mathbb{R}^m$ is B-differentiable at $x_0 \in U$ if and only if its component functions $f_i : U \to \mathbb{R}$, $i = 1, \ldots, m$, are B-differentiable at x_0.

An unrenounceable property for the practical use of the B-derivative is the existence of calculus rules which, in fact, are mere restatements of the classical rules. The most important calculus rule is the following chain rule.

Theorem 3.1.1. *If $U \subseteq \mathbb{R}^n$ and $V \subseteq \mathbb{R}^p$ are open sets and $f : U \to \mathbb{R}^m$ and $g : V \to \mathbb{R}^n$ are continuous and B-differentiable at the points $x_0 \in V$ and $g(x_0) \in U$, respectively, then the function $f \circ g$ is B-differentiable at x_0 and*

$$(f \circ g)'(x_0; y) = f'(g(x_0); g'(x_0; y)).$$

Proof. Define the functions

$$r_f(z) = f(g(x_0) + z) - f(g(x_0)) - f'(g(x_0); z),$$
$$r_g(y) = g(x_0 + y) - g(x_0) - g'(x_0; y)$$
$$r_{f \circ g}(y) = f(g(x_0 + y)) - f(g(x_0)) - f'(g(x_0); g'(x_0; y))$$

in suitable neighborhoods of the origin in $\mathbb{R}^n$ and $\mathbb{R}^p$, respectively. The definition of the B-derivative yields

$$\lim_{z \to 0} \frac{r_f(z)}{\|z\|} = 0 \quad \text{and} \quad \lim_{y \to 0} \frac{r_g(y)}{\|y\|} = 0. \tag{3.2}$$

Since $f'(g(x_0); g'(x_0; .))$ is positively homogeneous, the second part of Remark 3.1.1 shows that it suffices to prove that

$$\lim_{y \to 0} \frac{\|r_{f \circ g}(y)\|}{\|y\|} = 0. \tag{3.3}$$

As a consequence of the positive homogeneity of the B-derivatives, we obtain

$$
\begin{aligned}
\frac{\|r_{f \circ g}(y)\|}{\|y\|} &= \frac{\|f(g(x_0 + y)) - f(g(x_0)) - f'(g(x_0); g'(x_0; y))\|}{\|y\|} \\
&= \frac{\|f(g(x_0) + g'(x_0; y) + r_g(y)) - f(g(x_0)) - f'(g(x_0); g'(x_0; y))\|}{\|y\|} \\
&= \frac{\|r_f(g'(x_0; y) + r_g(y)) + f'(g(x_0); g'(x_0; y) + r_g(y)) - f'(g(x_0); g'(x_0; y))\|}{\|y\|} \\
&\leq \frac{\|r_f(g'(x_0; y) + r_g(y))\|}{\|y\|} \\
&\quad + \left\| f'\left(g(x_0); g'\left(x_0; \frac{y}{\|y\|}\right) + \frac{r_g(y)}{\|y\|}\right) - f'\left(g(x_0); g'\left(x_0; \frac{y}{\|y\|}\right)\right) \right\|.
\end{aligned}
$$

In order to complete the proof, it thus suffices to show that

$$\lim_{y \to 0} \frac{\|r_f(g'(x_0; y) + r_g(y))\|}{\|y\|} = 0 \tag{3.4}$$

and

$$\lim_{y \to 0} \left\| f' \left(g(x_0); g' \left(x_0; \frac{y}{\|y\|} \right) + \frac{r_g(y)}{\|y\|} \right) - f' \left(g(x_0); g' \left(x_0; \frac{y}{\|y\|} \right) \right) \right\| = 0.$$

(3.5)

The latter identity is a mere consequence of (3.2) and the continuity of the
B-derivatives $f'(g(x_0); .)$ and $g'(x_0; .)$ which these functions inherit from the
functions f and g (cf. Remark 3.1.1). To see (3.4), note that

$$\frac{\|r_f(g'(x_0; y) + r_g(y))\|}{\|y\|} \leq \frac{\|r_f(g'(x_0; y) + r_g(y))\|}{\|g'(x_0; y) + r_g(y)\|} \left(\left\| g' \left(x_0; \frac{y}{\|y\|} \right) \right\| + \frac{\|r_g(y)\|}{\|y\|} \right).$$

The continuity of $g'(x_0; .)$ and (3.2) implies the existence of a bound $M > 0$ with

$$\left(\left\| g' \left(x_0; \frac{y}{\|y\|} \right) \right\| + \frac{\|r_g(y)\|}{\|y\|} \right) \leq M$$

for every y in a neighborhood of the origin. Since $g'(x_0; y) + r_g(y)$ tends to zero
as y tends to zero, we may thus use (3.2) to deduce (3.4).   □

The following corollary is an immediate consequence of the chain rule and the
derivatives of the functions $f_1(x_1, x_2) = \alpha x_1 + \beta x_2$, $f_2(x_1, x_2) = x_1 x_2$, and
$f_3(x_1, x_2) = \frac{x_1}{x_2}$, where the last function is defined only if $x_2 \neq 0$.

Corollary 3.1.1. *Let $f, g : U \to \mathbb{R}$ be continuous B-differentiable functions and
$\alpha, \beta \in \mathbb{R}$. Then*

$$(\alpha f + \beta g)'(x_0; y) = \alpha f'(x_0; y) + \beta g'(x_0; y),$$

and

$$(fg)'(x_0; y) = g(x_0) f'(x_0; y) + f(x_0) g'(x_0; y).$$

If $g(x_0) \neq 0$, then

$$\left(\frac{f}{g} \right)'(x_0; y) = \frac{g(x_0) f'(x_0; y) - f(x_0) g'(x_0; y)}{g(x_0)^2}.$$

Example 3.1.1. As an illustration of the chain rule, we calculate the B-derivative of
the max-min function

$$f(x) = \max_{1 \leq i \leq l} \min_{j \in M_i} f_j(x),$$

where $M_i \subseteq \{1, \ldots, k\}$ for every $i = 1, \ldots, l$, and the functions $f_1, \ldots, f_k : U \to \mathbb{R}$
are C^1-functions defined on an open subset U of $\mathbb{R}^n$. Note first that the piecewise
linear function

$$\tilde{f}(v) = \max_{1 \leq i \leq l} \min_{j \in M_i} v_j$$

is B-differentiable in view of Proposition 2.2.6 and that the functions f_i are by assumption F-differentiable. Hence Theorem 3.1.1 shows that f is B-differentiable. To calculate the B-derivative of f at $x_0 \in U$, we use the chain rule of Theorem 3.1.1. For $z \in \mathbb{R}^l$, we define

$$h(z) = \max_{1 \le i \le l} z_i,$$

and for $i \in \{1, \dots, l\}$ and $v \in \mathbb{R}^k$, we define

$$g_i(v) = \min_{j \in M_i} v_j.$$

A direct application of the definition of the directional derivative yields

$$h'(z; u) = \max_{i \in I(z)} u_i,$$

$$g_i'(v; w) = \min_{j \in J_i(v)} w_j,$$

where $I(z) = \{i \,|\, h(z) = z_i\}$ and $J_i(v) = \{j \in M_i \,|\, g(v) = v_j\}$. Now we can use the chain rule to calculate the B-derivative of $\tilde{f}(v) = h(g_1(v), \dots, g_l(v))$ as

$$\tilde{f}'(v; w) = \max_{i \in \tilde{I}(v)} \min_{j \in \tilde{J}_i(v)} w_j,$$

where

$$\tilde{J}_i(v) = \left\{ j \in M_i \,\middle|\, \min_{p \in M_i} v_p = v_j \right\},$$

$$\tilde{I}(v) = \left\{ i \in \{1, \dots, k\} \,\middle|\, \tilde{J}_i(v) \ne \emptyset, \, \tilde{f}(v) = \min_{p \in \tilde{J}_i(v)} v_p \right\}.$$

A final application of the chain rule thus yields

$$f'(x; y) = \max_{i \in \tilde{I}(f(x))} \min_{j \in \tilde{J}_i(f(x))} \nabla f_j(x)^T y.$$

3.1.1   The B-Derivative of a Locally Lipschitz Continuous Function

An important class of nonsmooth functions are the locally Lipschitz continuous functions. Recall that a function $f : U \to \mathbb{R}^m$ is called *Lipschitz continuous* on $V \subseteq U$ if there exists a constant L such that

$$\|f(x) - f(y)\| \le L \|x - y\| \tag{3.6}$$

for every $x, y \in V$, i.e., the distance between two function values can be bounded by a linear function of the distance between the arguments. A constant L satisfying (3.6) is called a Lipschitz constant for f on V. The function f is called *locally Lipschitz continuous*, if every point $x \in U$ admits a neighborhood $V \subseteq U$ on which f is Lipschitz continuous. Note that a locally Lipschitz continuous function $f : U \to \mathbb{R}^m$ is Lipschitz continuous on every compact subset of U. Originally, S.M. Robinson introduced the B-derivative as the directional derivative of a locally Lipschitz continuous function. The reason that this definition is equivalent to ours in the case of a locally Lipschitz continuous function is a consequence of the following theorem.

Theorem 3.1.2. *Let $U \subseteq \mathbb{R}^n$ be an open set, and let $f : U \to \mathbb{R}^m$ be a locally Lipschitz continuous function which is directionally differentiable at $x_0 \in U$.*

1. *If L is a Lipschitz constant for f in a neighborhood of x_0, then L is a global Lipschitz constant for $f'(x_0; .)$.*
2. *The function f is B-differentiable at x_0.*

Proof. The first part is a direct consequence of the definition of the directional derivative. In fact,

$$\|f'(x_0; y) - f'(x_0; z)\| = \lim_{\substack{\alpha \to 0 \\ \alpha > 0}} \frac{1}{\alpha} \|f(x_0 + \alpha y) - f(x_0 + \alpha z)\|$$

$$\leq \lim_{\substack{\alpha \to 0 \\ \alpha > 0}} \frac{1}{\alpha} L \|\alpha y - \alpha z\|$$

$$= L \|y - z\|.$$

To prove part 2, we have to show that

$$\lim_{x \to x_0} \frac{\|f(x) - f(x_0) - f'(x_0; x - x_0)\|}{\|x - x_0\|} = 0.$$

Suppose this is not the case. Then there exists a sequence of vectors $y_m \in \mathbb{R}^n$ converging to the origin such that

$$\lim_{m \to \infty} \frac{\|f(x_0 + y_m) - f(x_0) - f'(x_0; y_m)\|}{\|y_m\|} \neq 0. \tag{3.7}$$

Since the unit sphere in $\mathbb{R}^n$ is compact, we may assume without loss of generality that the sequence $\frac{y_m}{\|y_m\|}$ converges to a vector $\bar{y}$ on the unit sphere, for otherwise we choose a suitable subsequence. Setting $\alpha_m = \|y_m\|$, we may use part 1 of the theorem and the positive homogeneity of the directional derivative to conclude that

$$0 \le \frac{\|f(x_0 + y_m) - f(x_0) - f'(x_0; y_m)\|}{\|y_m\|}$$

$$= \frac{1}{\alpha_m}\|f(x_0 + y_m) - f(x_0 + \alpha_m \bar{y}) + f'(x_0; \alpha_m \bar{y}) - f'(x_0; y_m)$$

$$+ f(x_0 + \alpha_m \bar{y}) - f(x_0) - f'(x_0; \alpha_m \bar{y})\|$$

$$\le \frac{1}{\alpha_m}\left(\|f(x_0 + y_m) - f(x_0 + \alpha_m \bar{y})\| + \|f'(x_0; \alpha_m \bar{y}) - f'(x_0; y_m)\|\right)$$

$$+ \frac{1}{\alpha_m}\|f(x_0 + \alpha_m \bar{y}) - f(x_0) - f'(x_0; \alpha_m \bar{y})\|$$

$$\le 2L \left\|\frac{y_m}{\alpha_m} - \bar{y}\right\| + \left\|\frac{1}{\alpha_m}(f(x_0 + \alpha_m \bar{y}) - f(x_0)) - f'(x_0; \bar{y})\right\|.$$

Since $\alpha_m = \|y_m\|$ is a nullsequence and $\frac{y_m}{\alpha_m}$ tends to $\bar{y}$, the definition of the directional derivative yields

$$\lim_{m \to \infty} \frac{\|f(x_0 + y_m) - f(x_0) - f'(x_0; y_m)\|}{\|y_m\|} = 0,$$

which contradicts (3.7) and thus proves the assertion. $\square$

An important tool for the applications of the classical differential calculus is the fundamental theorem of the calculus which relates derivatives and integrals. The following proposition yields an extension of this result to locally Lipschitz continuous B-differentiable functions.

Proposition 3.1.1. *Let $U \subseteq \mathbb{R}^n$ be an open convex set, $f : U \to \mathbb{R}^m$ be a locally Lipschitz continuous B-differentiable function, and let $x_0, x_1 \in U$. The function $\psi : [0,1] \to \mathbb{R}^m$ defined by $\psi(t) = f'(x_0 + t(x_1 - x_0); x_1 - x_0)$ is Lebesgue integrable and*

$$f(x_1) = f(x_0) + \int_0^1 f'(x_0 + t(x_1 - x_0); x_1 - x_0)dt.$$

Proof.   Note first that, in view of the chain rule, the function $\phi(t) = f(x_0 + t(x_1 - x_0))$ is locally Lipschitz continuous and B-differentiable and that

$$\phi'(t) = f'(x_0 + t(x_1 - x_0); x_1 - x_0) = \psi(t),$$

provided the derivative ϕ' exists at the point t. Since $\phi(0) = f(x_0)$ and $\phi(1) = f(x_1)$, it thus suffices to show that the ordinary derivative

$$\phi'(t) = \lim_{\tau \to t} \frac{\phi(\tau) - \phi(t)}{\tau - t}$$

exists for almost all $t \in [0, 1]$, that ϕ' is Lebesgue integrable, and that the identity

$$\phi(0) - \phi(1) = \int_0^1 \phi'(t) dt \tag{3.8}$$

holds. By a classical result of Lebesgue the function ϕ has the required properties if it is absolutely continuous on $[0, 1]$, i.e., if for every number $\varepsilon > 0$, there exists a number $\delta > 0$ such that for every $n \in \mathbb{N}$ and every collection of pairwise disjoint intervals $(\alpha_k, \beta_k) \subseteq [0, 1]$, $k = 1, \ldots, n$, with total length

$$\sum_{k=1}^n (\beta_k - \alpha_k) < \delta,$$

the inequality

$$\sum_{k=1}^n |\phi(\beta_k) - \phi(\alpha_k)| < \varepsilon$$

holds. To see this, note first that the locally Lipschitz continuous function ϕ is Lipschitz continuous on the compact set $[0, 1]$. Let $L > 0$ be a Lipschitz constant for the function ϕ on $[0, 1]$, $\varepsilon > 0$ be an arbitrarily chosen positive real number, and set $\delta = \frac{\varepsilon}{L}$. If $(\alpha_k, \beta_k) \subseteq [0, 1]$, $k = 1, \ldots, n$, is a finite collection of pairwise disjoint intervals with total length

$$\sum_{k=1}^n (\beta_k - \alpha_k) < \delta,$$

then the Lipschitz inequality yields

$$\sum_{k=1}^n |\phi(\beta_k) - \phi(\alpha_k)| \leq L \sum_{k=1}^n |\beta_k - \alpha_k| < \varepsilon,$$

which shows that ϕ is indeed absolutely continuous on $[0, 1]$.   □

3.1.2 Strongly B-Differentiable Functions

In applying the classical calculus, one often encounters situations in which the continuous dependence of the derivative on the argument plays an important role. The following proposition shows that for locally Lipschitz continuous B-differentiable functions this favorable situation can only occur if the function is indeed F-differentiable at the respective point.

Proposition 3.1.2. *Let U be an open neighborhood of $x_0 \in \mathbb{R}^n$ and let $f : U \to \mathbb{R}^m$ be locally Lipschitz continuous and B-differentiable. If for every fixed direction $y \in \mathbb{R}^n$ the function $f'(x; y)$ as a function of x is continuous at x_0, then f is F-differentiable at x_0.*

Proof. We have to prove that $f'(x_0; .)$ is linear. Since the directional derivative is positively homogeneous, it suffices to prove that it is additive as well. To see this, fix two nonvanishing directions $y, z \in \mathbb{R}^n$. In view of Proposition 3.1.1 and the continuity of the function $f'(.; z)$ at x_0 we obtain

$$0 \le \|f'(x_0; y + z) - f'(x_0; y) - f'(x_0; z)\|$$

$$= \lim_{\substack{\alpha \to 0 \\ \alpha > 0}} \left\| \frac{1}{\alpha} (f(x_0 + \alpha y + \alpha z) - f(x_0 + \alpha y)) - f'(x_0; z) \right\|$$

$$= \lim_{\substack{\alpha \to 0 \\ \alpha > 0}} \left\| \frac{1}{\alpha} \int_0^1 f'(x_0 + \alpha y + t\alpha z; \alpha z) dt - f'(x_0; z) \right\|$$

$$\le \lim_{\substack{\alpha \to 0 \\ \alpha > 0}} \int_0^1 \|f'(x_0 + \alpha y + t\alpha z; z) - f'(x_0; z)\| dt$$

$$= \int_0^1 \lim_{\substack{\alpha \to 0 \\ \alpha > 0}} \|f'(x_0 + \alpha y + t\alpha z; z) - f'(x_0; z)\| dt$$

$$= 0,$$

where the exchange of the limit and the integral is allowed as a consequence of Lebesgue's bounded convergence theorem which is applicable in view of the continuity of the function $f'(.; z)$ at the point x_0. In fact, since the function $f'(.; z)$ is continuous at x_0, we may choose an arbitrary bound $K > 0$. The continuity yields the existence of a real number $\delta > 0$ such that

$$\|f'(x_0 + v; z) - f'(x_0; z)\| \le K$$

for every $\|v\| < \delta$. If $0 < \alpha < \frac{\delta}{\|y\| + \|z\|}$ and $t \in [0, 1]$, then

$$\|\alpha y + t\alpha z\| \le \alpha(\|y\| + t\|z\|) \le \alpha(\|y\| + \|z\|) < \delta,$$

and hence

$$\|f'(x_0 + \alpha y + t\alpha z; z) - f'(x_0; z)\| \le K$$

for every $0 < \alpha < \frac{\delta}{\|y\| + \|z\|}$ and every $t \in [0, 1]$. $\qquad\square$

Note that the latter proposition does not claim that the function f is continuously F-differentiable in a neighborhood of the point x_0. In fact, the B-derivative of the function $f(x, y) = |x| y$ in a fixed direction $(v, w) \in \mathbb{R}^2$ is a continuous function of the argument (x, y) at the origin. Nevertheless, f is not F-differentiable in a neighborhood of the origin and thus not continuously F-differentiable at this point.

A concept which is somewhat weaker than continuous differentiability but considerably stronger than B-differentiability is the strong B-differentiability of a function. If $U \subseteq \mathbb{R}^n$ is an open set and $f, g : U \to \mathbb{R}^m$ are two functions, then g is called a *strong first-order approximation* of f at $x_0 \in U$ if $f(x_0) = g(x_0)$ and

$$\lim_{\substack{(y,z) \to (x_0, x_0) \\ y \neq z}} \frac{\|(f(y) - g(y)) - (f(z) - g(z))\|}{\|y - z\|} = 0.$$

A B-differentiable (F-differentiable) function $f : U \to \mathbb{R}^m$ is called *strongly B-differentiable* (F-differentiable) at x_0 if the function $g(x) = f(x_0) + f'(x_0; x - x_0)$ is a strong first-order approximation of f at x_0, i.e., if

$$\lim_{\substack{(y,z) \to (x_0, x_0) \\ y \neq z}} \frac{\|(f(y) - f(z)) - (f'(x_0; y - x_0) - f'(x_0; z - x_0))\|}{\|y - z\|} = 0.$$

Setting $z = x_0$, one readily verifies that a strong first order approximation of f is a first order approximation of f. We will see in the subsequent section that the strong approximation property can be used to generalize the classical inverse function theorem. However, before we present these results we show that strong B-differentiability is implied by the continuity of the B-derivative as a function of the base point.

Proposition 3.1.3. *Let $U \subseteq \mathbb{R}^n$ be an open neighborhood of x_0 and let $f : U \to \mathbb{R}^m$ be a locally Lipschitz continuous and B-differentiable. If for every $y \in \mathbb{R}^n$ the function $f'(.; y)$ as a function of x is continuous at x_0, then f is strongly F-differentiable at x_0.*

Proof. It has already been shown in Proposition 3.1.2 that f is F-differentiable at x_0. Consider two sequences of distinct vectors y_m, z_m which both converge to x_0. We have to prove that

$$\lim_{m \to \infty} \frac{\|(f(y_m) - f(z_m)) - (f'(x_0; y_m - x_0) - f'(x_0; z_m - x_0))\|}{\|y_m - z_m\|} = 0. \quad (3.9)$$

Set

$$v_m = \frac{y_m - z_m}{\|y_m - z_m\|}.$$

In view of Proposition 3.1.1 and the F-differentiability of f at x_0, the relation (3.9) is equivalent to

$$\lim_{m \to \infty} \left\| \int_0^1 f'(z_m + t(y_m - z_m); v_m) dt - \nabla f(x_0) v_m \right\| = 0. \quad (3.10)$$

Since it suffices to prove the latter relation for every convergent subsequence of the sequence of vectors v_m, we may assume that v_m converges to a vector $\bar{v}$ and thus obtain

$$\left\| \int_0^1 f'(z_m + t(y_m - z_m); v_m)dt - \nabla f(x_0)v_m \right\|$$

$$\leq \int_0^1 \| f'(z_m + t(y_m - z_m); v_m) - f'(z_m + t(y_m - z_m); \bar{v}) \|dt$$

$$+ \int_0^1 \| f'(z_m + t(y_m - z_m); \bar{v}) - f'(x_0; \bar{v}) \|dt$$

$$+ \| \nabla f(x_0)(\bar{v} - v_m) \|.$$

If L is a Lipschitz constant for f in some open convex neighborhood of x_0, then part 1 of Theorem 3.1.2 shows that for sufficiently large $m \in \mathbb{N}$ the first and third summand are bounded by $L\|v_m - \bar{v}\|$ and thus converge to zero. The convergence of the second summand is a consequence of the continuity of $f'(.; \bar{v})$ and Lebesgue's bounded convergence theorem, which, in view of the continuity assumption, implies that

$$\lim_{m \to \infty} \int_0^1 \| f'(z_m + t(y_m - z_m); \bar{v}) - f'(x_0; \bar{v}) \|dt$$

$$= \int_0^1 \lim_{m \to \infty} \| f'(z_m + t(y_m - z_m); \bar{v}) - f'(x_0; \bar{v}) \|dt = 0. \qquad \square$$

3.1.3   Comments and References

The notion of a Bouligand derivative has been introduced by Robinson in [64] for locally Lipschitz continuous functions. Most of the properties of B-derivatives can be found in the book [14] of Demyanov and Rubinov. For a comparison of the properties of various differential concepts, we refer to the article [75] of Shapiro. A proof of Theorem 3.1.2 can be found in the paper [51] of Pallaschke, Recht, and Urbański.

In the proofs of the Propositions 3.1.1–3.1.3 we have used some results from Lebesgue integration theory. For an account on the subject, we refer to the book [32] of Kolmogorov and Fomin. In particular, Lebesgue's bounded convergence theorem can be found in Chap. 8, Sect. 30.1, while Lebesgue's Theorem on the integrability of the derivative of an absolutely continuous function is presented in Chap. 9, Sect. 33.2.

Strongly B-differentiable functions have been introduced and studied by Robinson in [61].

3.2   Inverse and Implicit Function Theorems

We have seen in the introductory chapter that many problems in optimization and equilibrium theory can be formulated as the problem of finding a solution of an equation

$$f(x) = 0,$$

where $f : \mathbb{R}^n \to \mathbb{R}^n$ is a continuous, not necessarily differentiable function. An important question is how the solution of the problem changes if the data is perturbed. Assuming that the perturbation of the data is controlled by a finite dimensional parameter $y \in \mathbb{R}^m$, one can formulate this problem as a parameterized equation

$$f(x, y) = 0.$$

The question is how a given solution x_0 corresponding to an initial parameter vector y_0 varies as a function of the parameter. The preferable situation is the functional dependence of the solution on the parameter, i.e., to each parameter vector y there exists a unique solution vector $x(y)$ solving the equation $f(x, y) = 0$. However, such a global behavior cannot be expected in general. For instance the equation $f(x, y) = x^2 - y = 0$, has a unique solution if $y = 0$, two solutions for every $y > 0$ and no real solution for $y < 0$. The solutions in the (x, y)-space form the graph of the function $y(x) = x^2$. If the function f is slightly perturbed in such a way that the solution set of the new function $\tilde{f}$ in the (x, y)-space is close to the solution set of the old function f, then the new equation $\tilde{f}(x, y) = 0$ will also have two solutions for some parameter vectors y, i.e., the unpleasant situation persists under small perturbations of the function f. Since this situation is in a sense typical for nonlinear equations, we focus attention on local existence and uniqueness questions. If U and V are subsets of $\mathbb{R}^n$ and $\mathbb{R}^m$, respectively, $f : U \times V \to \mathbb{R}^n$ is a function, then the equation $f(x, y) = 0$ is said to determine an *implicit function* $x(y)$ at the point $(x_0, y_0) \in U \times V$ if $f(x_0, y_0) = 0$ and there exist neighborhoods $\tilde{U} \subseteq U$ and $\tilde{V} \subseteq V$ such that for every $y \in \tilde{V}$ the vector $x(y)$ is the unique solution of $f(x, y) = 0$ in the set $\tilde{U}$. Of course, one is not only interested in the existence and local uniqueness of a solution but also in the properties of the implicit function. Questions arising in this context are for instance:

- Is the implicit function continuous?
- Is it possible to estimate the distance of $x(y)$ from the original solution $x_0 = x(y_0)$ in terms of the distance from y_0 to y?
- Is the implicit function locally Lipschitz continuous and if so, how can we calculate a Lipschitz constant in a neighborhood of y_0?
- Is the implicit function differentiable at y_0 and if so, how can we calculate its derivative at the point y_0?

Statements which provide conditions for the existence of an implicit function are called implicit function theorems. For smooth functions, the following classical implicit function theorem holds.

Theorem 3.2.1. *Let $f : \mathbb{R}^n \times \mathbb{R}^m \to \mathbb{R}^n$ be a C^r-function, $r \geq 1$, and let $(x_0, y_0) \in \mathbb{R}^n \times \mathbb{R}^m$ be such that $f(x_0, y_0) = 0$. If $\nabla_x f(x_0, y_0)$ is a nonsingular matrix, then the equation $f(x, y) = 0$ determines an implicit function $x(y)$ at (x_0, y_0). Moreover the function x is a C^r-function and its Jacobian at the point y_0 is given by*

$$\nabla x(y_0) = -\nabla_x f(x_0, y_0)^{-1} \nabla_y f(x_0, y_0).$$

We will not prove this result since it is a special case of a more general implicit function theorem which will be stated and proved later. There are three properties which make the latter theorem extremely useful in applications:

- The implicit function x inherits the differentiability property of the function f.
- The condition is verifiable, provided the Jacobian of f is known.
- The Jacobian of x at y_0 can be easily calculated with the aid of the Jacobian of f at (x_0, y_0).

A straightforward generalization of the assumption of the implicit function theorem to nonsmooth functions would be the requirement that $f'((x_0, y_0); (., 0))$ is invertible, which in the C^1-case is equivalent to the nonsingularity of the restricted Jacobian $\nabla_x f(x_0, y_0)$. In the nonsmooth case, however, the latter condition does not guarantee that the equation $f(x, y)$ determines an implicit function at (x_0, y_0). The following example exhibits a particularly simple nonsmooth equation which satisfies the latter condition without determining an implicit function.

Example 3.2.1. Consider the parametric nonlinear complementarity problem

$$\mathbf{CP}(\varepsilon)\ x_1, x_2 \geq 0,$$
$$f_1(x_1, x_2) = \frac{1}{2}(x_1 + x_2) - \frac{1}{4}(x_1 + x_2 + \varepsilon)^2 \geq 0$$
$$f_2(x_1, x_2) = \frac{1}{2}(x_1 + x_2) - \frac{1}{4}(x_1 + x_2 + \varepsilon)^2 \geq 0$$
$$f(x)^T x = 0.$$

A reformulation of $CP(\varepsilon)$ as a parametric nonsmooth equation can be done via

$$F(x_1, x_2, \varepsilon) = \begin{pmatrix} \min\left\{x_1, \frac{1}{2}(x_1 + x_2) - \frac{1}{4}(x_1 + x_2 + \varepsilon)^2\right\} \\ \min\left\{x_2, \frac{1}{2}(x_1 + x_2) - \frac{1}{4}(x_1 + x_2 + \varepsilon)^2\right\} \end{pmatrix} = \begin{pmatrix} 0 \\ 0 \end{pmatrix}.$$

The vector $(x_1, x_2) = (0, 0)$ solves $CP(0)$. Moreover, using the theory of piecewise affine functions or direct calculation, one easily verifies that

$$F'((0, 0, 0); (y_1, y_2, 0)) = \begin{pmatrix} \min\left\{y_1, \frac{1}{2}(y_1 + y_2)\right\} \\ \min\left\{y_2, \frac{1}{2}(y_1 + y_2)\right\} \end{pmatrix}$$

is a homeomorphism. In fact, its inverse is the function

$$G(z_1, z_2) = \begin{pmatrix} \max\{z_1, 2z_1 - z_2\} \\ \max\{z_2, 2z_2 - z_1\} \end{pmatrix}.$$

Nevertheless, the equation $F(x_1, x_2, \varepsilon) = 0$ does not determine an implicit function at $(0, 0, 0)$. In fact, define the function $\phi : (-\infty, \frac{1}{2}) \to \mathbb{R}$ by

$$\phi(\varepsilon) = 1 - \varepsilon - \sqrt{1 - 2\varepsilon}.$$

Note that ϕ is a C^∞-function and

$$\phi'(\varepsilon) = -1 + \frac{1}{\sqrt{1 - 2\varepsilon}},$$

$$\phi''(\varepsilon) = \left(\frac{1}{\sqrt{1 - 2\varepsilon}} \right)^3.$$

Hence ϕ is a strictly convex function and $\varepsilon = 0$ is its unique minimizer; thus

$$\phi(\varepsilon) > \phi(0) = 0$$

for every nonvanishing $\varepsilon < \frac{1}{2}$. Using the latter inequality, a direct calculation yields

$$F(\phi(\varepsilon), 0, \varepsilon) = \begin{pmatrix} \min\{\phi(\varepsilon), 0\} \\ \min\{0, 0\} \end{pmatrix} = F(0, \phi(\varepsilon), \varepsilon) = \begin{pmatrix} \min\{0, 0\} \\ \min\{\phi(\varepsilon), 0\} \end{pmatrix} = \begin{pmatrix} 0 \\ 0 \end{pmatrix}$$

for every $\varepsilon \in (-\infty, \frac{1}{2})$. Moreover, $\phi(\varepsilon) = 0$ if and only if $\varepsilon = 0$. Hence the equation $F(x_1, x_2, \varepsilon) = 0$ has at least two distinct solutions for every nonvanishing $\varepsilon \in (-\infty, \frac{1}{2})$ and thus does not determine an implicit function in a neighborhood of $(0, 0, 0)$.

The latter example is the starting point for our study of parametric nonsmooth equations. Our aim is to provide conditions which ensure that the invertibility of the restricted B-derivative $f'((x_0, y_0); (., 0))$ implies the existence of an implicit function $x(y)$ determined by the equation $f(x, y) = 0$ in a neighborhood of a solution (x_0, y_0).

A particularly easy form of a parametric equation is the right-hand side perturbation

$$f(x) = y,$$

of an equation $f(x) = 0$, where $x, y \in \mathbb{R}^n$. In this case, the equation $f(x) - y = 0$ determines an implicit function at (x_0, y_0) if and only if the function f has a local inverse function at x_0. The additional requirement that the inverse function is continuous leads to the local homeomorphism problem for the function f at the point x_0. We have already introduced the notion of a homeomorphism in connection with piecewise affine functions. We say that a function $f : U \to V$ mapping a subset U of $\mathbb{R}^n$ into a subset V of $\mathbb{R}^m$ is *invertible* if the equation $f(x) = y$ has a unique solution $x = f^{-1}(y) \in U$ for every $y \in V$. The function f^{-1} is called the *inverse function* of f. Thus a continuous function is a homeomorphism if and only if it is invertible and open, i.e., the images of open sets in U are open sets in V. In the following we will be mainly concerned with *Lipschitz homeomorphisms*, i.e., invertible functions $f : U \to V$ with the property that f and f^{-1} are Lipschitz continuous on U and V, respectively. There are also local versions of the latter notions. The function $f : U \to V$ is called invertible at $x_0 \in U$ if there exist neighborhoods $\tilde{U} \subseteq U$ of x_0 and $\tilde{V} \subseteq V$ of $f(x_0)$ such that the equation $f(x) = y$ has a unique solution $x = f_{x_0}^{-1}(y) \in \tilde{U}$ for every $y \in \tilde{V}$. The function $f_{x_0}^{-1} : \tilde{V} \to \tilde{U}$ is called a local inverse of f at x_0. If f and $f_{x_0}^{-1}$ are Lipschitz continuous on $\tilde{U}$ and $\tilde{V}$, then f is said to be *Lipschitz-invertible at x_0*. A function which is Lipschitz-invertible at x_0 is sometimes also called a *local Lipschitz homeomorphisms* at x_0.

There are three celebrated theorems of L.E.J. Brouwer which are related to the homeomorphism problem.

Open mapping theorem: *If $U \subseteq \mathbb{R}^n$ is open and $f : U \to \mathbb{R}^n$ is continuous and injective, then f is open.*

Invariance of domain: *If U, V are subsets of $\mathbb{R}^n$ and $f : U \to V$ is a homeomorphism, then boundary points of U are mapped onto boundary points of V.*

Invariance of dimension: *If $U \subseteq \mathbb{R}^n$ and $V \subseteq \mathbb{R}^m$ are open sets and $f : U \to V$ is a homeomorphism, then $n = m$.*

Note that the last statement can be directly deduced from the second result by embedding the space of smaller dimension into the space of larger dimension. The latter results belong to the heart of combinatorial topology. We omit the proofs since they are quite involved and can be found in standard textbooks on the subject (cf. Sect. 3.2.4).

Verifying whether a given function is a homeomorphism or not is usually very difficult. In the case of a C^r-function, however, the following inverse function theorem provides a tool to check whether a given function is locally Lipschitz invertible or not.

Theorem 3.2.2. *If $U \subseteq \mathbb{R}^n$ is open and $f : U \to \mathbb{R}^n$ is a C^r-function, $r \geq 1$, then f is locally Lipschitz invertible at x_0 if and only if the Jacobian $\nabla f(x_0)$ is nonsingular. Moreover, the local inverse of f is a C^r-function in a neighborhood of $f(x_0)$.*

We will not prove this theorem here since a generalization is proved later on for strongly B-differentiable functions. As in the case of the implicit function theorem, the question arises, whether the invertibility of the B-derivative of a nonsmooth function is a sufficient criterion for the function to be invertible. The following modification of Example 3.2.1 shows that this is not the case in general.

Example 3.2.2. Consider the function

$$H(z_1, z_2) = \begin{pmatrix} \min\left\{ z_1, \frac{1}{2}(z_1 + z_2) - \frac{1}{4}(z_1 + z_2)^2 \right\} \\ \min\left\{ z_2, \frac{1}{2}(z_1 + z_2) - \frac{1}{4}(z_1 + z_2)^2 \right\} \end{pmatrix}.$$

An application of the chain rule yields

$$H'((0,0);(y_1, y_2) = \begin{pmatrix} \min\left\{ y_1, \frac{1}{2}(y_1 + y_2) \right\} \\ \min\left\{ y_2, \frac{1}{2}(y_1 + y_2) \right\} \end{pmatrix}.$$

The B-derivative of H at the origin is a homeomorphism and its inverse function is given by

$$G(z_1, z_2) = \begin{pmatrix} \max\{z_1, 2z_1 - z_2\} \\ \max\{z_2, 2z_2 - z_1\} \end{pmatrix}.$$

Nevertheless, H is not invertible at the origin. To see this, let F be the function defined in Example 3.2.1, i.e.,

$$F(x_1, x_2, \varepsilon) = \begin{pmatrix} \min\left\{ x_1, \frac{1}{2}(x_1 + x_2) - \frac{1}{4}(x_1 + x_2 + \varepsilon)^2 \right\} \\ \min\left\{ x_2, \frac{1}{2}(x_1 + x_2) - \frac{1}{4}(x_1 + x_2 + \varepsilon)^2 \right\} \end{pmatrix} = \begin{pmatrix} 0 \\ 0 \end{pmatrix}.$$

A direct calculation shows that

$$H\left(x_1 + \frac{1}{2}\varepsilon, x_2 + \frac{1}{2}\varepsilon \right) - \begin{pmatrix} \frac{1}{2}\varepsilon \\ \frac{1}{2}\varepsilon \end{pmatrix} = F(x_1, x_2, \varepsilon).$$

Choosing $\varepsilon \in (-\infty, \frac{1}{2})$, Example 3.2.1 yields the solutions $(\phi(\varepsilon) + \frac{1}{2}\varepsilon, \frac{1}{2}\varepsilon)$ and $(\frac{1}{2}\varepsilon, \phi(\varepsilon) + \frac{1}{2}\varepsilon)$ of the equation

$$H(z_1, z_2) = \begin{pmatrix} \frac{1}{2}\varepsilon \\ \frac{1}{2}\varepsilon \end{pmatrix}.$$

The solutions are distinct for every nonvanishing $\varepsilon < \frac{1}{2}$, i.e., H is not injective in a neighborhood of the origin.

Inverse function theorems, being special cases of implicit function theorems, are not as special as it might seem at first glance. In fact there is a classical approach to create implicit function theorems from inverse function theorems.

Lemma 3.2.1. *Let U, V be subsets of $\mathbb{R}^n$ and $\mathbb{R}^m$, respectively, $f : U \times V \to \mathbb{R}^n$ be a function, and let $(x_0, y_0) \in U \times V$, $z_0 = f(x_0, y_0)$. The equation $f(x, y) - z = 0$ determines an implicit function $x(y, z)$ at (x_0, y_0) if and only if the function $g(x, y) = (f(x, y), y)$ has a local inverse at (x_0, y_0). In fact, if $g^{-1}(v, w) = (g_x^{-1}(v, w), g_y^{-1}(v, w)) \in U \times V$, then $x(y, z) = g_x^{-1}(z, y)$.*

Proof. The equation $f(x, y) = z$ holds if and only if $g(x, y) = (z, y)$. If, on the one hand, the function g has a local inverse function at (x_0, y_0), then there exists a neighborhood $W \subseteq U \times V$ of (x_0, y_0) which is bijectively mapped onto a neighborhood $g(W)$ of (z_0, y_0). Hence the equation $f(x, y) = z$ has a unique solution $x = g_x^{-1}(z, y)$ for every $(y, z) \in g(W)$. If, on the other hand, there exist neighborhoods $\tilde{U}$ of x_0, $\tilde{V}$ of y_0, and $\tilde{W}$ of z_0) such that the equation $f(x, y) - z = 0$ has a unique solution $x(y, z) \in \tilde{U}$ for every $(y, z) \in \tilde{V} \times \tilde{W}$, then $g(x, y)$ maps $\tilde{U} \times \tilde{V}$ bijectively onto $\tilde{W} \times \tilde{V}$ and its inverse is given by $g^{-1}(z, y) = (x(y, z), y)$. $\square$

Remark 3.2.1. The latter lemma shows that the equation $f(x, y) = 0$ determines an implicit function $x(y)$ at (x_0, y_0), provided that $g(x, y) = (f(x, y), y)$ has a local inverse at (x_0, y_0) and that in this case the implicit function $x(y)$ is given by $x(y) = g_x^{-1}(y, 0)$. However, the converse of this statement is not true in general. A trivial counterexample is provided by function $f(x, y) = x^2$. The equation $f(x, y) = 0$ determines the implicit function $x(y) = 0$ at $(0, 0)$. However, the equation $f(x, y) = z$ has no solution for $z < 0$ and thus g does not have an inverse function at the origin.

The classical implicit function theorem can be proved with the aid of the corresponding inverse function theorem using the latter lemma. This shows that every inverse function theorem yields an implicit function theorem and we can thus focus attention to inverse function theorems.

We close this section with an elementary but very useful lemma which characterizes local Lipschitz homeomorphisms.

Lemma 3.2.2. *A locally Lipschitz continuous function $f : U \to \mathbb{R}^n$ is a local Lipschitz homeomorphism at x_0 if and only if there exists a neighborhood $V \subseteq U$ of x_0 and a constant $l > 0$ such that $\| f(x_1) - f(x_2) \| \geq l \|x_1 - x_2\|$ for every $x_1, x_2 \in V$.*

Proof. Recall that f is a local Lipschitz homeomorphism at x_0 if and only if there exist neighborhoods V of x_0 and W of $f(x_0)$ and a locally Lipschitz continuous mapping $g : W \to V$ such that for every $x \in V$ and $y \in W$ the equation $f(x) = y$ holds if and only if $x = g(y)$. Suppose on the one hand that f is a local Lipschitz homeomorphism at x_0. If we choose $l > 0$ in such a way that $\frac{1}{l}$ is a Lipschitz constant of g in W, then

$$\|x_1 - x_2\| = \|g(y_1) - g(y_2)\| \le \frac{1}{l}\|y_1 - y_2\| = \frac{1}{l}\|f(x_1) - f(x_2)\|,$$

which proves the inequality.

Suppose on the other hand that the inequality holds for all x_1, x_2 in a neighborhood V of x_0 and let $W = f(V)$. The validity of the inequality implies that $f(x_1) = f(x_2)$ if and only if $x_1 = x_2$, i.e., f is injective. Hence there exists an inverse function $g : W \to V$ with the property that $f(x) = y$ for $x \in V$ and $y \in W$ if and only if $x = g(y)$. Moreover, if $x_1 = g(y_1)$ and $x_2 = g(y_2)$, then

$$\|g(y_1) - g(y_2)\| = \|x_1 - x_2\| \le \frac{1}{l}\|f(x_1) - f(x_2)\| = \frac{1}{l}\|y_1 - y_2\|,$$

whence g maps V homeomorphically onto W. Since V is a neighborhood of x_0, the invariance of domain theorem shows that W is a neighborhood of $f(x_0)$ and thus f is a local Lipschitz homeomorphism at x_0.   □

3.2.1   B-Derivatives of Local Lipschitz Homeomorphisms

We have seen in Example 3.2.2 that the Lipschitz invertibility of the B-derivative of a nonsmooth function does not necessarily imply the local Lipschitz invertibility of the function. The aim of this section is to study the relation between the Lipschitz invertibility of the B-derivative and the Lipschitz invertibility of the function. We begin with an elementary observation about positively homogeneous functions.

Proposition 3.2.1. *Let $f : \mathbb{R}^n \to \mathbb{R}^n$ be a positively homogeneous function.*

1. *If f is invertible at the origin, then f is invertible on $\mathbb{R}^n$ and the inverse f^{-1} is positively homogeneous.*
2. *If f is a local Lipschitz homeomorphism at the origin, then f is a Lipschitz homeomorphism.*

Proof. 1. Let $u \in \mathbb{R}^n$ be an arbitrary vector. The positive homogeneity of f implies that $f(x) = u$ if and only if $f(\lambda x) = \lambda u$ for every $\lambda > 0$. Since f is invertible at the origin, there exists a $\lambda > 0$ such that $f(x) = \lambda u$ has a solution x_λ. Hence $f(\frac{1}{\lambda} x_\lambda) = u$, which proves the surjectivity. Moreover, if $f(x) = f(y)$, then $f(\lambda x) = f(\lambda y)$ for every $\lambda \ge 0$. The local injectivity of f yields $\lambda x = \lambda y$ for some $\lambda > 0$ and thus $x = y$, which shows that f is injective. The positive homogeneity of f^{-1} is easily seen since $f^{-1}(\lambda u)$ is the unique solution to the equation $f(x) = \lambda u$. Hence $\frac{1}{\lambda} f^{-1}(\lambda u) = f^{-1}(u)$.

2. Using the first statement, it suffices to prove that every Lipschitz constant L of a positively homogeneous function f in a neighborhood U of the origin is a global Lipschitz constant for f. To see this, let $x, y \in \mathbb{R}^n$ and choose $\lambda > 0$ such that $\lambda x, \lambda y \in U$. Then

$$\| f(x) - f(y)\| = \frac{1}{\lambda}\|f(\lambda x) - f(\lambda y)\| \le \frac{1}{\lambda}L\|\lambda x - \lambda y\| = L\|x - y\|. \qquad \square$$

The following theorem illuminates the relation between the local Lipschitz invertibility of a B-differentiable function and of its B-derivative.

Theorem 3.2.3. *Let $U \subseteq \mathbb{R}^n$ be open and $f : U \to \mathbb{R}^n$ be a continuous function which is B-differentiable at $x_0 \in U$.*

1. *If f is a local Lipschitz homeomorphism at x_0, then its B-derivative $f'(x_0; .)$ is a Lipschitz homeomorphism. Moreover, the local inverse function f^{-1} is B-differentiable at $f(x_0)$ and its B-derivative is the inverse of the function $f'(x_0; .)$.*
2. *If $f'(x_0; y) = 0$ implies $y = 0$, then there exists a neighborhood $V \subseteq U$ of x_0 such that the only solution of the equation $f(x) = f(x_0)$ in the set V is $x = x_0$.*
3. *If the B-derivative $f'(x_0; .)$ is a Lipschitz homeomorphism, then $f(x_0)$ is an interior point of $f(V)$ for every neighborhood $V \subseteq U$ of x_0.*

Proof. Considering the function $\tilde{f}(x) = f(x - x_0) - f(x_0)$ instead of f if necessary, we may assume without loss of generality that $x_0 = f(x_0) = 0$.

1. Since f is a local Lipschitz homeomorphism at x_0, Lemma 3.2.2 yields the existence of a constant $l \ge 0$ and a neighborhood $\tilde{U}$ of x_0 such that

$$\|f(x) - f(y)\| \ge l\|x - y\|$$

for every $x, y \in \tilde{U}$. Hence

$$\|f'(x_0; v) - f'(x_0; w)\| = \left\| \lim_{\substack{\alpha \to 0 \\ \alpha > 0}} \frac{1}{\alpha}(f(x_0 + \alpha v) - f(x_0 + \alpha w))\right\|$$

$$\ge \lim_{\substack{\alpha \to 0 \\ \alpha > 0}} l\|\alpha v - \alpha w\|$$

$$= l\|v - w\|,$$

which, in view of Lemma 3.2.2, shows that $f'(x_0; .)$ is a local Lipschitz homeomorphism at the origin and thus a Lipschitz homeomorphism.

To see the second part of the statement, let $h(y) = f'(x_0; y)$ and let L be a common Lipschitz constant of h^{-1} in $\mathbb{R}^n$ and of f^{-1} in a neighborhood V of $f(x_0)$. Define

$$y(v) = f^{-1}(f(x_0) + v) - x_0$$

for $v \in V - \{f(x_0)\}$. Since $f^{-1} : V \to f^{-1}(V)$ is a homeomorphism, we obtain

$$y(v) = 0 \text{ if and only if } v = 0.$$

Moreover, a direct application of the definition of $y(v)$ yields

$$\|y(v)\| = \|f^{-1}(f(x_0) + v) - f^{-1}(f(x_0))\| \leq L\|v\|,$$

$$v = f(x_0 + y(v)) - f(x_0),$$

and

$$\lim_{\|v\| \to 0} \|y(v)\| = 0. \tag{3.11}$$

With these properties, one easily verifies that the following chain of inequalities holds for every nonvanishing $v \in V - \{f(x_0)\}$:

$$0 \leq \frac{\|f^{-1}(f(x_0) + v) - f^{-1}(f(x_0)) - h^{-1}(v)\|}{\|v\|}$$

$$\leq \frac{L\|h\big(f^{-1}(f(x_0 + v) - x_0)\big) - v\|}{\|v\|}$$

$$\leq \frac{L\|f'(x_0; y(v)) - f(x_0 + y(v)) + f(x_0)\|}{\|v\|}$$

$$\leq \frac{L^2\|f(x_0 + y(v)) - f(x_0) - f'(x_0; y(v))\|}{\|y(v)\|}$$

If $\|v\|$ tends to zero, then the B-differentiability of f together with (3.11) implies that the last term tends to zero and thus all the other terms tend to zero as well. In particular,

$$\lim_{\|v\| \to 0} \frac{\|f^{-1}(f(x_0) + v) - f^{-1}(f(x_0)) - h^{-1}(v)\|}{\|v\|} = 0.$$

Since the inverse of a positively homogeneous function is again positively homogeneous, the latter identity shows that h^{-1} is indeed the B-derivative of f^{-1} at the point $f(x_0)$.
2. We have to show that there exists a positive real number δ such that x_0 is the only solution to the equation $f(x) = f(x_0)$ in the ball with radius δ. By assumption the equation $f'(x_0; y) = 0$ has the only solution $y = 0$. Hence the positive homogeneity of the B-derivative yields

$$\|f'(x_0; y)\| \geq \varepsilon\|y\|$$

for every $y \in \mathbb{R}^n$, where $\varepsilon > 0$ is the minimum of the continuous function $\|f'(x_0; .)\|$ on the compact unit sphere. Due to the approximation property of the B-derivative, we can find a real number $\delta > 0$ such that

$$\|f(x_0 + y) - f(x_0) - f'(x_0; y)\| < \varepsilon\|y\|$$

for every $y \in \mathbb{R}^n$ with $0 < \|y\| < \delta$. Combining both inequalities, we obtain

$$\|f(x_0 + y) - f(x_0) - f'(x_0; y)\| < \|f'(x_0; y)\|.$$

The triangle inequality immediately implies

$$0 < \|f(x_0 + y) - f(x_0)\|$$

for every $y \in \mathbb{R}^n$ with $0 < \|y\| < \delta$, i.e., x_0 is the only solution of the equation $f(x) = f(x_0)$ in the ball around x_0 with radius δ.
3. We have to prove that for every sufficiently small $\varepsilon > 0$ there exists a real number $\delta > 0$ such that for every z with $\|z\| \le \delta$ the equation $f(x_0 + y) = f(x_0) + z$ has a solution y with $\|y\| \le \varepsilon$. Setting $h(y) = f'(x_0; y)$, the equation $f(x_0 + y) = f(x_0) + z$ is equivalent to the fixed-point equation

$$g_z(y) = y - h^{-1}(f(x_0 + y) - f(x_0) - z) = y.$$

If L is a Lipschitz constant of the function h^{-1}, then we obtain

$$\begin{aligned}
\|g_z(y)\| &= \|y - h^{-1}(f(x_0 + y) - f(x_0) - z)\| \\
&\le \|h^{-1}(h(y)) - h^{-1}(f(x_0 + y) - f(x_0))\| \\
&\quad + \|h^{-1}(f(x_0 + y) - f(x_0)) - h^{-1}(f(x_0 + y) - f(x_0) - z)\| \\
&\le L(\|f'(x_0; y) - f(x_0 + y) + f(x_0)\| + \|z\|).
\end{aligned}$$

We have to show that for every $\varepsilon > 0$ there exists a real number $\delta > 0$ such that for every $\|z\| \le \delta$ the function g_z has a fixed point y_z with $\|y_z\| \le \varepsilon$. Brouwer's fixed-point theorem states that every continuous function which maps a convex compact set into itself has a fixed point. Since g_z is continuous, it thus suffices to provide positive numbers δ and $\hat{\varepsilon} \le \varepsilon$ such that $\|g_z(y)\| \le \hat{\varepsilon}$ for every $\|z\| \le \delta$ and every $\|y\| \le \hat{\varepsilon}$. The B-differentiability of f shows that there exists a number $\tilde{\varepsilon}$ such that

$$\|f(x_0 + y) - f(x_0) - f'(x_0; y)\| \le \frac{1}{2L}\|y\|$$

for every $\|y\| \le \tilde{\varepsilon}$. Choosing $\hat{\varepsilon} = \min\{\tilde{\varepsilon}, \varepsilon\}$ and $\delta = \frac{\hat{\varepsilon}}{2L}$, we conclude from the above estimation of $\|g_z(y)\|$ that

$$\|g_z(y)\| \le L \left(\frac{1}{2L} \|y\| + \frac{\hat{\varepsilon}}{2L} \right) \le \hat{\varepsilon}$$

for every $\|y\| \le \hat{\varepsilon}$ and every $\|z\| \le \delta$, which proves the assertion. □

3.2.2 An Inverse Function Theorem

The following theorem can be readily used to generalize the classical inverse function theorem to the class of locally Lipschitz continuous strongly B-differentiable functions.

Theorem 3.2.4. *Let $U \subseteq \mathbb{R}^n$ be an open set and $f, g : U \to \mathbb{R}^n$ be locally Lipschitz continuous functions. If g is a strong first order approximation of f at x_0, then f is a local Lipschitz homeomorphism at x_0 if and only if g is a local Lipschitz homeomorphism at x_0.*

Proof. Suppose g is a local Lipschitz homeomorphism at x_0. Then there exists a constant $l > 0$ such that

$$\|g(y) - g(z)\| \ge l \|y - z\|$$

for every x, y in a neighborhood $V \subseteq U$ of x_0. Moreover, the strong approximation property shows that for every $\varepsilon > 0$ there exists a real number $\delta > 0$ such that

$$\|(f(y) - f(z)) - (g(y) - g(z))\| \le \varepsilon \|y - z\|,$$

whenever $\|y - x_0\| \le \delta$ and $\|z - x_0\| \le \delta$. Hence

$$\begin{aligned}\|f(y) - f(z)\| &= \|(g(y) - g(z)) - [(f(z) - f(y)) - (g(z) - g(y))]\| \\ &\ge \|g(y) - g(z)\| - \|(f(z) - f(y)) - (g(z) - g(y))\| \\ &\ge l \|y - z\| - \varepsilon \|y - z\|\end{aligned}$$

Choosing $\varepsilon < l$, we thus obtain a positive constant $\tilde{l} = l - \varepsilon$ such that

$$\|f(y) - f(z)\| \ge \tilde{l} \|y - z\|$$

for every $y, z \in V$ with $\|y - x_0\| \le \delta$ and $\|z - x_0\| \le \delta$. Lemma 3.2.2 thus shows that f is a local Lipschitz homeomorphism at x_0. Since g is a strong approximation of f at x_0 if and only if f is a strong approximation of g at x_0, we have thus completed the proof. □

The following corollary is an immediate consequence of the definition of a strong B-derivative.

Corollary 3.2.1. *If the locally Lipschitz continuous function $f : U \to \mathbb{R}^n$ has a strong B-derivative at x_0, then f is a local Lipschitz homeomorphism at x_0 if and only if its B-derivative is a local Lipschitz homeomorphism at the origin.*

3.2.3   Hadamard's Theorem

We close this chapter with an extension of Hadamard's Theorem to locally Lipschitz continuous B-differentiable functions. The original version of this theorem states that a C^1-function $f : \mathbb{R}^n \to \mathbb{R}^n$ which is a local diffeomorphism at every point $x \in \mathbb{R}^n$ is a global diffeomorphism provided the set of all matrices $\nabla f(x)^{-1}$, $x \in \mathbb{R}^n$, is bounded.

The basis for a proof of the latter theorem is the continuation property. A continuous mapping $f : \mathbb{R}^n \to \mathbb{R}^n$ has the *continuation property* for a path $q :$ $[0,1] \to \mathbb{R}^n$ if for every $a \in (0,1]$ and every continuous function $p : [0,a) \to \mathbb{R}^n$ satisfying $f(p(t)) = q(t)$ for every $t \in [0,a)$ there exists a number $p(a)$ such that $p(t)$ tends to a $p(a)$ if t tends to a from below. Note that the continuity of f and q implies that $f(p(a)) = q(a)$. The following lemma is a variant of the Homotopy Lifting Theorem 2.3.4.

Lemma 3.2.3 ((cf. [49], (5.3.4))).   *Let $f : \mathbb{R}^n \to \mathbb{R}^n$ be a local homeomorphism, $r : [0,1] \to \mathbb{R}^n$ be a path, and $H : [0,1] \times [0,1] \to \mathbb{R}^n$ be a continuous function such that $H(s,0) = f(r(s))$ for all $s \in [0,1]$. If for each $s \in [0,1]$ the function f has the continuation property for the path $H(s,.)$, then there exists a unique continuous mapping $\hat{H} : [0,1] \times [0,1] \to \mathbb{R}^n$ such that $\hat{H}(.,0) = r$ and $f \circ \hat{H} = H$. Moreover, if $H(s,1) = H(0,t) = H(1,t)$ for every $s,t \in [0,1]$, then $r(0) = r(1)$.*

With the aid of the latter lemma, it is possible to prove the following generalization of Hadamard's theorem.

Theorem 3.2.5.   *Let $f : \mathbb{R}^n \to \mathbb{R}^n$ be a B-differentiable local Lipschitz homeomorphism at each point $x \in \mathbb{R}^n$. If there exists a constant $l > 0$ such that*

$$\| f'(x;y) \| \geq l \| y \| \text{ for every } x, y \in \mathbb{R}^n, \tag{3.12}$$

then f maps $\mathbb{R}^n$ homeomorphically onto $\mathbb{R}^n$.

Proof. The proof is based on the fact that f has the continuation property for every locally Lipschitz continuous B-differentiable path $q : [0,1] \to \mathbb{R}^n$. To see this, let $a \in (0,1]$ and $p : [0,a) \to \mathbb{R}^n$ be a continuous function with $f(p(t)) = q(t)$ for every $t \in [0,a)$. Since f is a local homeomorphism, the identity

$$p(t) = f^{-1}(q(t)) \tag{3.13}$$

holds in a neighborhood of $\bar{t}$, where f^{-1} is a local inverse of f at $p(\bar{t})$. Since q as well as f^{-1} are locally Lipschitz continuous and B-differentiable, so is the function p. In order to see that

$$\lim_{\substack{t \to a \\ t < a}} p(t) = p(a) \tag{3.14}$$

exists, we first apply the chain rule and part 1 of Theorem 3.2.3, and conclude from (3.13) and the assumption of the theorem that

$$\|p'(t; 1)\| \le \frac{1}{l}\|q'(t; 1)\|. \tag{3.15}$$

for every $t \in (0, a)$. Since any locally Lipschitz continuous function is globally Lipschitz continuous on compact sets, there is a global Lipschitz constant L for q and hence

$$\|q'(v; w)\| \le L\|w\| \tag{3.16}$$

for every $v \in (0, a)$ and every $w \in \mathbb{R}$. Combining the inequalities (3.15) and (3.16), we thus obtain

$$\|p'(t; 1)\| \le \frac{L}{l} \tag{3.17}$$

for every $t \in (0, a)$. Hence we conclude from Proposition 3.1.1 that

$$\|p(t_1) - p(t_0)\| \le \frac{L}{l}|t_1 - t_0| \tag{3.18}$$

for every $t_0, t_1 \in (0, a)$ and thus, in view of the Cauchy convergence criterion, the limit (3.14) exists which establishes the continuation property of f for Lipschitz continuous B-differentiable paths.

In order to establish the surjectivity of f, fix a vector $x_0 \in \mathbb{R}^n$, set $y_0 = f(x_0)$, and choose an arbitrary vector $y_1 \in \mathbb{R}^n$. Define a path $q : [0, 1] \to \mathbb{R}^n$ by $q(t) = ty_1 + (1-t)y_0$. Setting $H(s, t) = q(t)$, we may apply Lemma 3.2.3 and obtain the existence of a path $p : [0, 1] \to \mathbb{R}^n$ with $p(0) = x_0$ and $f \circ p = q$. In particular, $f(p(1)) = q(1) = y_1$ which proves that f is surjective.

To see that f is injective, let $x_0, x_1 \in \mathbb{R}^n$ with $f(x_0) = f(x_1)$. Define the path $r(s) = (1-s)x_0 + sx_1$ and consider the mapping $H : [0, 1] \times [0, 1] \to \mathbb{R}^n$ defined by

$$H(s, t) = tf(x_0) + (1-t)f(sx_1 + (1-s)x_0).$$

Clearly H satisfies the assumptions of Lemma 3.2.3 and $H(s, 1) = H(0, t) = H(1, t)$ for every $s, t \in [0, 1]$. Hence $x_0 = x_1$ and thus f is injective.

The openness of f is a direct consequence of the local homeomorphism property (cf. Proposition 2.3.9).   □

Recall that by Theorem 2.3.6 a local homeomorphism $f : \mathbb{R}^n \to \mathbb{R}^n$ is a homeo-morphism if and only if f is closed. It would be interesting to have a direct proof of the fact that a function f satisfying the assumptions of Hadamard's Theorem is a closed mapping without referring to Lemma 3.2.3.

3.2.4   Comments and References

For a proof of the invariance of domain theorem, we refer to the classical monograph [1] of Alexandroff and Hopf or to Spanier's book [78]. A proof of the open mapping theorem based on degree theory can be found in Deimling's book [12].

Lemma 3.2.2 and part 1 of Theorem 3.2.3 are due to Kummer (cf. [35], Lemma 1, and [37], Lemma 2). The local surjectivity result of Theorem 3.2.3 was essentially already known to Alexandroff and Hopf (cf. [1], p. 477–478, in particular "Bemerkung I" and "Bemerkung II" p. 478). The proof given here is based on an idea of Halkin (cf. [24, 39]). For an account on Brouwer's fixed-point theorem we refer to [49]. As a reference for the inverse function theorem for strongly B-differentiable functions we mention Robinson's articles [61, 65]. For an account on the continuation property, in particular for a proof of Lemma 3.2.3, we refer to the monograph [49] of Ortega and Rheinbold. The proof of Theorem 3.2.5 is close to the proof of the classical theorem given in Schwartz's book [74] (cf. [74], Theorem 1.22, and [49], (5.3.10)). For a more general result in the context of strongly B-differentiable functions we refer to Theorem 3.3 in Robinson's paper [65].

3.3   Appendix: Inverse Function Theorems of Clarke and Kummer

There are some inverse function theorems close in spirit to the classical inverse function theorem for differentiable functions which are applicable to general locally Lipschitz continuous functions. They all incorporate one or another generalization of the Jacobian of a differentiable function. The most important of these concepts is F.H. Clarke's notion of the *generalized Jacobian* of a locally Lipschitz continuous function $f : \mathbb{R}^n \to \mathbb{R}^m$ which is defined by

$$\partial f(x_0) = \operatorname{conv} \bigcap_{\varepsilon > 0} \operatorname{cl}\{\nabla f(x) \,|\, \|x - x_0\| < \varepsilon, x \in \Omega\},$$

where $\Omega \subseteq \mathbb{R}^n$ is the set of all points where f is F-differentiable (cf. [9]). In other words, the generalized Jacobian is the convex hull of all limit points of Jacobian sequences $\nabla f(x_k)$, where the sequence of base-points $x_k \in \Omega$ converges to x_0. The basis of the differential theory of locally Lipschitz continuous functions is Rademacher's Theorem which ensures that the set of points where a locally Lipschitz continuous function fails to be F-differentiable is a Lebesgue nullset. This classical result is used to prove the following proposition.

Proposition 3.3.1 ([9], Proposition 2.6.2). *If* $f : \mathbb{R}^n \to \mathbb{R}^m$ *is a locally Lipschitz continuous function, then* $\partial f(x_0)$ *is a nonempty convex compact set of* $m \times n$-*matrices.*

Clarke's generalized Jacobian was the first analytic device to deal with nonsmooth equations and is still very useful for many purposes. In particular, it can be used to formulate an inverse function theorem.

Theorem 3.3.1 (Clarke's Inverse Function Theorem [9], Theorem 7.1.1). *Let* $f : \mathbb{R}^n \to \mathbb{R}^n$ *be a locally Lipschitz continuous function. If all matrices in the generalized Jacobian* $\partial f(x_0)$ *of* f *at the point* $x_0 \in \mathbb{R}^n$ *are nonsingular, then* f *is a local Lipschitz homeomorphism at* x_0.

The main drawback of Clarke's inverse function theorem is the fact that it involves the nonsingularity of a generally infinite number of matrices. Moreover, it is not as close to the classical theorem has one might think. In the C^1-case, the nonsingularity of the Jacobian matrix is a necessary and sufficient criterion for a function to be a local Lipschitz homeomorphism. Kummer gave a simple example, which shows that this is not the case for Clarke's condition (cf. [36]). He provided necessary and sufficient conditions for a locally Lipschitz continuous function to be a local Lipschitz homeomorphism in terms of a generalized directional derivative

$$\Delta f(x_0; y) = \bigcap_{\varepsilon > 0} \mathrm{cl} \left\{ \frac{f(x + \alpha y) - f(x)}{\alpha} \,\middle|\, \|x - x_0\| < \varepsilon, 0 < \alpha < \varepsilon \right\}.$$

In other words, $\Delta f(x_0; y)$ consists of all limit points of sequences of vectors

$$z_n = \frac{f(x_n + \alpha_n y) - f(x_n)}{\alpha_n},$$

where the sequence of vectors x_n tends to x_0 and the sequence of real α_n tends to zero from above.

Theorem 3.3.2 (Kummer's Inverse Function Theorem [35], Theorem 1). *A locally Lipschitz continuous function* $f : \mathbb{R}^n \to \mathbb{R}^n$ *is a local Lipschitz homeomorphism at* $x_0 \in \mathbb{R}^n$ *if and only if* $0 \notin \Delta f(x_0; y)$ *for every nonvanishing vector* $y \in \mathbb{R}^n$.

Despite the theoretical benefits of the inverse function theorems of F.H. Clarke and B. Kummer, they share the drawback that their assumptions are not very handy in applications. In particular, one cannot hope to find a finite algorithm for their verification. This is due to the fact that in general locally Lipschitz continuous functions do not admit a sufficiently simple local approximation scheme.

Chapter 4
Piecewise Differentiable Functions

4.1 Basic Notions and Properties

We start with some basic notions. Let U be a subset of $\mathbb{R}^n$ and let $f_i : U \to \mathbb{R}^m$, $i = 1, \ldots, k$ be a collection of continuous functions. A function $f : U \to \mathbb{R}^m$ is said to be a *continuous selection* of the functions $f_1, \ldots, f_k$ on the set $O \subseteq U$ if f is continuous on O and $f(x) \in \{f_1(x), \ldots, f_k(x)\}$ for every $x \in O$. A function $f : U \to \mathbb{R}^m$ defined on an open set $U \subseteq \mathbb{R}^n$ is called a PC^r-*function*, for some $r \in \mathbb{N} \cup \{\infty\}$ if for every $x_0 \in U$ there exists an open neighborhood $O \subseteq U$ and a finite number of C^r-functions $f_i : O \to \mathbb{R}^m$, $i = 1, \ldots, k$, such that f is a continuous selection of $f_1, \ldots, f_k$ on O. A set of C^r-functions $f_i : O \to \mathbb{R}^m$, $i = 1, \ldots, k$, defined on an open neighborhood $O \subseteq U$ of x_0 is called a set of *selection functions* for the PC^r-function f at x_0 if $f(x) \in \{f_1(x), \ldots, f_k(x)\}$ for every $x \in O$.

If we say that a function f is a continuous selection of the functions $f_1, \ldots, f_k$, we tacitly assume that the domains of definition of all functions coincide and that f is a continuous selection of the functions f_i on the common domain. A PC^r-function is thus a function which is everywhere locally a continuous selection of C^r-functions. PC^1-functions are also called *piecewise differentiable* functions.

Examples of piecewise differentiable functions can be found in the introductory chapter. In fact, the nonsmooth equations presented there all involve piecewise differentiable functions if the data functions of the original problems are sufficiently smooth.

The superposition $g \circ f$ of two PC^r-functions f and g is again a PC^r-function. In fact, if f is a continuous selection of the C^r-functions $f_1, \ldots, f_k$ in the open neighborhood O of x_0 and g is a continuous selection of the C^r-functions $g_1, \ldots, g_l$ in the open neighborhood V of $f(x_0)$, then $g \circ f$ is a continuous selection of the C^r-functions $g_j \circ f_i$, $i \in \{1, \ldots, k\}$, $j \in \{1, \ldots, l\}$ in the open neighborhood $f^{-1}(V) \cap O$ of x_0. In particular, a function $f : U \to \mathbb{R}^m$ is a PC^r-function if and only if each component function f_i, $i = 1, \ldots, m$, is PC^r. Moreover, scalar multiples and finite sums of PC^r-functions are again PC^r, and, in the case of real-valued functions, the product as well as the pointwise maximum or

S. Scholtes, *Introduction to Piecewise Differentiable Equations*, SpringerBriefs in Optimization, DOI 10.1007/978-1-4614-4340-7_4, © Stefan Scholtes 2012

minimum of two PC^r-functions is PC^r. In particular, the maximum and minimum operations readily allow the construction of piecewise differentiable functions. More sophisticated examples will be considered later.

Given a set of selection functions $f_1, \ldots, f_k$ for a PC^r-function f at a point x_0, we define the *active index set* at the point x_0 by

$$I_f(x_0) = \{i \in \{1, \ldots, k\} \mid f(x_0) = f_i(x_0)\}. \tag{4.1}$$

The selection functions f_i, $i \in I_f(x_0)$, are called *active selection functions* at x_0. Since the functions f_i are C^r and a fortiori continuous, the relation $f(x) \neq f_i(x)$ persists in a neighborhood U_{x_0} of x_0, whence $I_f(x) \subseteq I_f(x_0)$ for every $x \in U_{x_0}$. So, if we are only interested in local properties of the function f in a neighborhood of x_0, which will be mainly the case throughout this chapter, we may assume that f is a continuous selection of the functions $f_i, i \in I_f(x_0)$. In some cases it is reasonable to exclude superfluous indices from $I_f(x_0)$. Consider for instance the following function of two variables:

$$f(x, y) = (\max\{x, y\}, \min\{x, y\}).$$

A set of selection functions for such max-min-type PC^r-functions can be constructed by combining all selection functions of the components. In our example this yields the functions $(x, x), (x, y), (y, x)$, and (y, y). All these selection functions are active at the origin. Nevertheless, the functions (x, x) and (y, y) are not essential for building the function f. In fact they are only active on the principal diagonal, which is a lower-dimensional set. This observation leads us to the introduction of the set of *essentially active indices*

$$I_f^e(x_0) = \{i \in \{1, \ldots, k\} \mid x_0 \in \mathrm{cl}(\mathrm{int}\{x \in U \mid f(x) = f_i(x)\})\},$$

where we assume that the C^r-functions $f_1, \ldots, f_k$ form a set of selection functions for the PC^r-function f at x_0. A selection function f_i is called *essentially active* at x_0 if $i \in I_f^e(x_0)$.

Our first result shows a piecewise differentiable function is everywhere locally a continuous selection of essentially active selection functions.

Proposition 4.1.1. *If $f : U \to \mathbb{R}^m$ is a PC^r-function and $x_0 \in U$, then there exists a collection of selection functions for f at x_0 which are all essentially active.*

Proof. We will prove the assertion by induction on the number k of selection functions. If f admits a single selection function at x_0, then this selection function coincides with f in a neighborhood of x_0 and is thus essentially active. So suppose the assertion holds for every PC^r-function defined on U which admits a collection of less than k selection functions at x_0 and let $f_1, \ldots, f_k : O \to \mathbb{R}^m$ be a set of selection functions corresponding to the PC^r-function f at x_0. Assuming that not all selection functions are essentially active, we will reduce the number of selection functions for f at x_0. By definition an index i is essentially active at x_0 if there exists a sequence x_n converging to x_0 with $x_n \in \text{int}\{x \in O | f(x) = f_i(x)\}$. Thus, if the index $i \in \{1, \ldots, k\}$ is not essentially active at x_0, then there exists a neighborhood V of x_0 with

$$V \cap \text{int}\{x \in O | f(x) = f_i(x)\} = \emptyset. \tag{4.2}$$

Moreover, if $\bar{x} \in O$ with $f(\bar{x}) = f_i(\bar{x})$ and $f(\bar{x}) \neq f_j(\bar{x})$ for every $i \neq j$, then these relations persists in a neighborhood of $\bar{x}$ due to the continuity of f and f_j and the assumption that f is a continuous selection of the functions $f_1, \ldots, f_k$. We thus obtain $\bar{x} \in \text{int}\{x \in O | f(x) = f_i(x)\}$ and, in view of (4.2), $\bar{x} \notin V$. Hence for every $x \in V$ there exists an index $j \in \{1, \ldots, k\} \backslash \{i\}$ with $f(x) = f_j(x)$, and thus f is a continuous selection of the functions $f_i, i \in I_f(x_0) \backslash \{i\}$ on the open neighborhood V of x_0. An induction argument thus completes the proof. $\qquad \square$

4.1.1 Local Lipschitz Continuity

In this section, we prove the important fact that a piecewise differentiable function is locally Lipschitz continuous. First of all, it is not difficult to verify that every C^1-function is locally Lipschitz continuous. In fact, if $f : U \to \mathbb{R}^m$ is C^1 and $O \subseteq U$ is a compact neighborhood of x_0, then the continuity of the gradient mapping shows that there exists a number L such that

$$\|\nabla f(x)\| \leq L$$

for every $x \in O$. Choosing a convex subneighborhood $V \subseteq O$ of x_0, we thus obtain

$$\|f(x) - f(y)\| = \left\| \int_0^1 \nabla f(x + t(y - x))^T (y - x) dt \right\|$$
$$\leq \int_0^1 \|\nabla f(x + t(y - x))\| \|y - x\| dt$$
$$\leq L \|y - x\|$$

for every $x, y \in V$. We will show in this section that every continuous selection of locally Lipschitz continuous functions is locally Lipschitz continuous, thus implying this property for the special case of piecewise differentiable functions. In order to prove this result, we make use of the following elementary lemma.

Lemma 4.1.1. *Let $\phi : [0, 1] \to \mathbb{R}^m$ be a continuous function and let $A_1, \ldots, A_l \subseteq \mathbb{R}^m$ be closed sets. If $\phi([0, 1]) \subseteq \cup_{i=1}^l A_i$, then there exist real numbers $0 = t_0 \leq \cdots \leq t_l = 1$ and corresponding indices $i_0, \ldots, i_{l-1}$ such that*

$$\{\phi(t_j), \phi(t_{j+1})\} \subseteq A_{i_j}$$

for every $j = 0, \ldots, l - 1$

Proof. The proof of this assertion is done by induction on the number l. The claim is certainly true if $l = 1$. In fact, in this case we may choose $t_0 = 0$, $t_1 = 1$, and $i_0 = 1$. Suppose the assertion of the lemma holds for all curves $\tilde{\phi} : [0, 1] \to \mathbb{R}^m$ and at most $l - 1$ closed sets $\tilde{A}_i$, and suppose the function $\phi : [0, 1] \to \mathbb{R}^m$ and the sets $A_1, \ldots, A_l$ satisfy the assumptions of the lemma. Let $i_0 \in \{1, \ldots, l\}$ be an index with $\phi(0) \in A_{i_0}$ and define

$$t_1 = \sup\{t \in [0, 1] | \phi(t) \in A_{i_0}\}.$$

The continuity of ϕ and the closedness of A_{i_0} imply that $\phi(t_1) \in A_{i_0}$. If on the one hand $t_1 = 1$, then the assertion holds with $t_1 = t_2 = \cdots = t_l$ and $i_0 = i_1 = \cdots = i_{l-1}$. On the other hand, if $t_1 < 1$, then the maximality of t_1 shows that $\phi(t) \notin A_{i_0}$ for every $t_1 < t \le 1$ and the continuity of ϕ yields $\phi(t_1) \in A_{i_1}$ for some index $i_1 \ne i_0$. Hence the continuous function $\tilde{\phi}(t) = \phi(t_1 + t(1 - t_1))$ defined on the interval $[0, 1]$ passes through the $l - 1$ closed sets A_i, $i \in \{1, \ldots, k\}\backslash\{i_0\}$. The induction assumption thus yields a sequence of numbers $0 = \tilde{t}_0 \le \cdots \le \tilde{t}_{l-1} = 1$ and indices $\tilde{i}_0, \ldots, \tilde{i}_{l-2}$ such that

$$\{\phi(\tilde{t}_j), \phi(\tilde{t}_{j+1})\} \subseteq A_{\tilde{i}_j}$$

for every $j = 0, \ldots, l - 2$. Setting $t_j = t_1 + \tilde{t}_{j-1}(1 - t_1)$ for $j = 1, \ldots, l$ and $i_j = \tilde{i}_{j-1}$ for $j = 1, \ldots, l - 1$ thus yields the assertion. $\qquad\square$

The following result is easily established with the aid of the latter lemma.

Proposition 4.1.2. *Let $V \subseteq \mathbb{R}^n$ be a convex set and $f_i : V \to \mathbb{R}^m$, $i = 1, \ldots, l$, be Lipschitz continuous on V with Lipschitz constants L_i, $i = 1, \ldots, l$. If $f : V \to \mathbb{R}^m$ is continuous and $f(x) \in \{f_1(x), \ldots, f_l(x)\}$, then f is Lipschitz continuous on V with Lipschitz constant $L = \max\{L_1, \ldots, L_l\}$.*

Proof. Fix $x, y \in V$ and define $\phi(t) = (x + t(y - x))$ for $t \in [0, 1]$ and

$$A_i = \{\phi(t) | f(\phi(t)) = f_i(\phi(t)), t \in [0, 1]\}$$

for $i = 1, \ldots, l$. Since the functions f and ϕ are continuous, the sets A_i are closed. Hence Lemma 4.1.1 yields the existence of real numbers $0 = t_0 < t_1 < \cdots < t_l = 1$ and indices $i_0, \ldots, i_{l-1}$ such that $\{\phi(t_j), \phi(t_{j+1})\} \subseteq A_{i_j}$ for every $j = 0, \ldots, l-1$. Due to the definition of the sets A_i, we thus obtain

$$f(x + t_j(y - x)) = f_{i_j}(x + t_j(y - x))$$
$$f(x + t_{j+1}(y - x)) = f_{i_j}(x + t_{j+1}(y - x)) \qquad (4.3)$$

for every $j = 0,\ldots,l-1$. Since

$$f(x) - f(y) = \sum_{j=0}^{l-1} f(x + t_j(y-x)) - f(x + t_{j+1}(y-x)),$$

we deduce from (4.3) and the assumptions of the proposition that

$$
\begin{aligned}
\|f(x) - f(y)\| &= \left\| \sum_{j=0}^{l-1} f_{i_j}(x + t_j(y-x)) - f_{i_j}(x + t_{j+1}(y-x)) \right\| \\
&\leq \sum_{j=0}^{l-1} \|f_{i_j}(x + t_j(y-x)) - f_{i_j}(x + t_{j+1}(y-x))\| \\
&\leq \sum_{j=0}^{l-1} L_{i_j} \|x - y\|(t_{j+1} - t_j) \\
&\leq \max\{L_1,\ldots,L_k\}\|x - y\| \sum_{j=0}^{l-1}(t_{j+1} - t_j) \\
&= L\|x - y\|. \qquad\qquad \square
\end{aligned}
$$

The main result of this section is a direct consequence of the latter proposition and the fact that C^1-functions are locally Lipschitz continuous. It is formulated in the following corollary.

Corollary 4.1.1. *Every piecewise differentiable function is locally Lipschitz continuous. A Lipschitz constant in a neighborhood of x_0 is given by the maximum of the Lipschitz constants of the selection functions.*

4.1.2 B-Differentiability

The B-differentiability of piecewise differentiable functions is of fundamental importance for our study of this function class.

Proposition 4.1.3. *Let $f: U \to \mathbb{R}^m$ be PC^1-function, let $f_1,\ldots,f_k: O \to \mathbb{R}^m$ be a collection of C^1-selection functions for f at $x_0 \in O \subseteq U$.*

1. *The function f is B-differentiable at x_0 and its B-derivative is a continuous selection of the F-derivatives of the essentially active selection functions, i.e.,*

$$f'(x_0; y) \in \{\nabla f_i(x_0)y \,|\, i \in I_f^e(x_0)\}.$$

2. *In addition, the function is directionally continuously differentiable in the following sense:*

$$\lim_{y \to 0} \frac{f'(x_0 + y; y) - f'(x_0; y)}{\|y\|}) = 0.$$

Proof. Let $\phi(t) = f(x_0 + ty)$ and $\phi_i(t) = f_i(x_0 + ty)$, $i \in I_f^e(x_0)$, be defined on a suitably small interval $(-\delta, \delta)$. Then ϕ is a PC^1 function with $\phi(0) = \phi_i(0)$ and for every $t \in (-\delta, \delta)$ we have $\phi(t) \in \{\phi_i(t) \mid i \in I_f^e(x_0)\}$. Since the functions ϕ_i are continuously differentiable, the relation

$$\phi_i(t) = \phi_i(0) + \int_0^t \phi_i'(s) ds$$

holds and hence, reducing δ if necessary, we deduce from $\phi_i'(0) \neq \phi_j'(0)$ that $\phi_i(t) \neq \phi_j(t)$ for every $t \in (0, \delta)$. Therefore there exists an index set $I \subseteq I_f^e(x_0)$ such that $\phi_i'(0) = \phi_j'(0)$ for every $i, j \in I$ and $\phi(t) \in \{\phi_i(t) \mid i \in I\}$ for every $t \in (0, \delta)$. Considering subsequences $\{\alpha_n\}$ corresponding to different active functions, one easily verifies that

$$\phi'(0; 1) = \lim_{\substack{\alpha \to 0 \\ \alpha > 0}} \frac{1}{\alpha}(\phi(\alpha) - \phi(0)) = \phi_i'(0) \tag{4.4}$$

for every $i \in I$. Hence ϕ is directionally differentiable at the origin and

$$\phi'(0; 1) \in \{\phi_i'(0) \mid i \in I(0)\}.$$

This proves the first part of the proposition. To see the second part, let y_n be a null sequence. Since f is PC^1 there is an index sequence $i_n \in I_f^e(x_0)$ with $f(x_0 + y_n) = f_{i_n}(x_0 + y_n)$ for all sufficiently large n. Notice that also $f(x_0) = f_{i_n}(x_0)$ as $i_n \in I_f^e(x_0)$. It suffices to prove the equation for all subsequences of y_n with a fixed index $i_n = i$. Since f_i is continuously differentiable we have

$$\lim_{n \to \infty} \frac{f_i(x_0 + y_n - y_n) - (f_i(x_0 + y_n) + \nabla f_i(x_0 + y_n)(-y_n))}{\|y_n\|} = 0,$$

and the B-differentiability of f implies

$$\lim_{n \to \infty} \frac{f(x_0 + y_n) - (f(x_0) + f'(x_0; y_n))}{\|y_n\|} = 0.$$

Combining these two equations we obtain

$$\lim_{n \to \infty} \frac{f'(x_0 + y_n; y_n) - f'(x_0; y_n)}{\|y_n\|}$$

$$= \lim_{n \to \infty} \frac{f_i(x_0 + y_n - y_n) - (f_i(x_0 + y_n) + \nabla f_i(x_0 + y_n)(-y_n))}{\|y_n\|}$$

$$+ \lim_{n \to \infty} \frac{f(x_0 + y_n) - (f(x_0) + f'(x_0; y_n))}{\|y_n\|}$$

$$= 0. \qquad \qquad \square$$

Note that the second statement of the latter proposition cannot be generalized to locally Lipschitz continuous B-differentiable functions which can be seen by considering the function

$$f(x) = \begin{cases} x^2 \sin \dfrac{1}{x} & \text{if } x \neq 0, \\ 0 & \text{if } x = 0 \end{cases}$$

The function f is F-differentiable on $\mathbb{R}$ with

$$f'(x) = \begin{cases} 2x \sin \dfrac{1}{x} + \cos \dfrac{1}{x} & \text{if } x \neq 0, \\ 0 & \text{if } x = 0. \end{cases}$$

Moreover, since the derivative f' is locally bounded, the mean value theorem implies the local Lipschitz continuity of f. Nevertheless, setting $t_n = \frac{1}{2n\pi}$, we obtain

$$f'(0 + t_n; 1) = f'(t_n) = 1,$$

which does not converge to $f'(0; 1) = 0$ as n tends to ∞.

4.1.3 Strong B-Differentiability

Beside the continuously differentiable functions, there is an important class of strongly B-differentiable but not necessarily F-differentiable PC^1-functions. This is the class of PC^1-functions for which the points where the function fails to be continuously differentiable is locally contained in a conical set. Recall that a set $C \subseteq \mathbb{R}^n$ is called a cone if $\lambda c \in C$ for every $c \in C$ and every $\lambda > 0$.

Proposition 4.1.4. *Let $f : U \to \mathbb{R}^m$ be a PC^1-function and let $f_1, \ldots, f_k : O \to \mathbb{R}^m$ be a collection of C^1-selection functions for f at $x_0 \in O \subseteq U$. If there exists a neighborhood $V \subseteq O$ of x_0 and a collection $C_1, \ldots, C_l \subseteq \mathbb{R}^n$ of closed cones such that*

1. $\mathbb{R}^n = \cup_{i=1}^l C_i$
2. *For every $i \in \{1, \ldots, l\}$ there exists an index $p(i) \in \{1, \ldots, k\}$ such that $f(x) = f_{p(i)}(x)$ for every $x \in V \cap (\{x_0\} + C_i)$*

then f is strongly B-differentiable at x_0.

Proof. A direct application of the definition shows that the function f is strongly B-differentiable at x_0 if and only if the function $\tilde{f}$ defined on $U - \{x_0\}$ by $\tilde{f}(x) = f(x - x_0)$ is strongly B-differentiable at the origin. Since the function $\tilde{f}$ satisfies all assumptions of the proposition for the origin instead of x_0, it suffices to prove the claim for the function $\tilde{f}$. Hence we may assume without loss of generality that $x_0 = 0$, which simplifies the exposition of the proof. Moreover, taking a smaller neighborhood if necessary, we may assume that V is convex.

Our first observation is that the directional derivatives $f'(0; y)$ and $f'_{p(i)}(0; y)$ coincide in the direction $y \in C_i$. In fact, if $y \in C_i$, then $f(\alpha y) = f_{p(i)}(\alpha y)$ for every sufficiently small $\alpha > 0$, which shows that

$$f'(y) = \nabla f_{p(i)}(0) y \tag{4.5}$$

for every $y \in C_i$. The proof is now carried out in two steps:

1. In the first step we show that for any two points $y, z \in V$ there exist real numbers $0 = t_0 \leq \cdots \leq t_l = 1$ and indices $q_0, \ldots, q_{l-1} \in \{1, \ldots, k\}$ such that the vectors $v_j = y + t_j(z - y)$ satisfy the identities

$$f(v_j) - f'(0; v_j) = f_{q_j}(v_j) - f'_{q_j}(0; v_j)$$
$$f(v_{j+1}) - f'(0; v_{j+1}) = f_{q_j}(v_{j+1}) - f'_{q_j}(0; v_{j+1}), \tag{4.6}$$

where $j \in \{0, \ldots, l - 1\}$. Setting $\phi(t) = y + t(z - y)$, Lemma 4.1.1 yields real numbers $0 = t_0 \leq \cdots \leq t_l = 1$ and indices $i_0, \ldots, i_{l-1}$ such that $\{\phi(t_j), \phi(t_{j+1})\} \subseteq C_{i_j}$ for every $j = 0, \ldots, l - 1$. Setting $v_j = \phi(t_j) = y + t_j(z - y)$ for $j = 0, \ldots, l$, we thus obtain

$$\{v_j, v_{j+1}\} \subseteq C_{i_j}. \tag{4.7}$$

Since $y, z \in V$ and V is convex, the vectors v_j are all contained in V and since the set C_{i_j} are cones, we conclude from (4.7) that $f(\alpha v_j) = f_{p(i_j)}(\alpha v_j)$ and $f(\alpha v_{j+1}) = f_{p(i_j)}(\alpha v_{j+1})$ for every $\alpha \in [0, 1]$. Setting $q_j = p(i_j)$ and applying the definition of the directional derivative thus prove the identities (4.6).

2. Now we are ready to prove the assertion of the proposition. Let $\varepsilon > 0$ be arbitrarily chosen. Proposition 3.1.3 yields the strong F-differentiability of the selection functions f_p at the point $x_0 = 0$. Hence there exists a number $\delta > 0$ such that

$$\|f_p(u) - f_p(w) - f'_p(0; u) + f'_p(0; w)\| \leq \varepsilon \|u - w\| \tag{4.8}$$

for any two points $u, w \in V$ with $\|u\| < \delta$ and $\|w\| < \delta$ and every $p \in \{1, \ldots, k\}$. Let $y, z \in V$ with $\|y\| < \delta$ and $\|z\| < \delta$. According to part 1 there exist reals $0 = t_0 \leq \ldots \leq t_l = 1$ and indices $q_0, \ldots, q_{l-1} \in \{1, \ldots, k\}$ such that the points $v_j = y + t_j(z - y)$, $j = 0, \ldots, l-1$, satisfy (4.6). Since the vectors v_j are contained in the line segment joining y and z, we obtain $\|v_j\| < \delta$ for every $j = 0, \ldots, l$. Since $v_0 = y$ and $v_l = z$, we further obtain

$$f(y) - f(z) - f'(0; y) + f'(0; z) = \sum_{j=0}^{l-1} \left(f(v_j) - f(v_{j+1}) - f'(0; v_j) + f'(0; v_{j+1}) \right).$$

Hence (4.6) and (4.8) yield

$$\|f(y) - f(z) - f'(0; y) + f'(0; z)\|$$

$$= \left\| \sum_{j=0}^{l-1} f(v_j) - f(v_{j+1}) - f'(0; v_j) + f'(0; v_{j+1}) \right\|$$

$$\leq \sum_{j=0}^{l-1} \| f_{q_j}(v_j) - f_{q_j}(v_{j+1}) - f'_{q_j}(0; v_j) + f'_{q_j}(0; v_{j+1}) \|$$

$$\leq \sum_{j=0}^{l-1} \varepsilon \|v_j - v_{j+1}\|$$

$$= \sum_{j=0}^{l-1} \varepsilon \|y - z\|(t_{j+1} - t_j)$$

$$= \varepsilon \|y - z\|.$$

Since $\varepsilon > 0$ was chosen arbitrary and y, z where arbitrary points in V with $\|y\| < \delta$ and $\|z\| < \delta$, we conclude that f is strongly B-differentiable at x_0. $\quad\square$

4.1.4 Continuous Differentiability

We close the analysis of the differentiability properties of piecewise differentiable functions with the proof that a PC^r-function is r-times continuously differentiable in a generic point set. As a first result, we mention Rademacher's Theorem which states that a locally Lipschitz continuous function defined on a subset of $\mathbb{R}^n$ is F-differentiable at all points outside of a Lebesgue nullset. Since a PC^r-function is locally Lipschitz continuous (cf. Corollary 4.1.1), this statement carries on to

PC^r-functions. However, as far as higher order derivatives are concerned, the Lipschitz property is of no use. In fact, if $g : \mathbb{R} \to \mathbb{R}$ is a continuous function, then the function

$$f(t) = \int_0^t g(s)ds$$

is continuously differentiable and a fortiori locally Lipschitz and $f'(t) = g(t)$ for every $t \in \mathbb{R}$. Taking as g a continuous nowhere differentiable function thus yields a locally Lipschitz continuous function f which is nowhere twice differentiable. Here, PC^r-functions behave much nicer. In fact, the following result shows that a PC^r-function is r-times continuously differentiable in a generic point set.

Proposition 4.1.5. *Let U and be an open subset of $\mathbb{R}^n$, and $f : U \to \mathbb{R}^m$ be a PC^r-function. There exists an open and dense subset $U' \subseteq U$ such that f is r-times continuously differentiable on U'.*

Proof. Let $f_1, \ldots, f_k : O \to \mathbb{R}^m$ be a set of C^r-selection functions for f at a point $x_0 \in U$. Clearly f is r-times continuously differentiable on the set

$$O' = \bigcup_{i=1}^k \text{int}\{x \in O | f(x) = f_i(x)\}$$

which is a finite union of open sets and thus itself open. Let us show that O' is a dense subset of O. To see this, we fix a point $\bar{x} \in O$ and construct a sequence $x_n \in O'$ which converges to $\bar{x}$. Since O is open, there exists a real number $\bar{\alpha}$ such that every closed ball $B(\bar{x}, \alpha)$ with center $\bar{x}$ and radius $0 < \alpha \le \bar{\alpha}$ is a subset of O. Since f is a continuous selection of the functions f_i, $i = 1, \ldots, k$, the set O is representable as

$$O = \bigcup_{i=1}^k \{x \in O | f(x) = f_i(x)\}.$$

Thus we obtain

$$B(\bar{x}, \alpha) = \bigcup_{i=1}^k \{x \in B(\bar{x}, \alpha) | f(x) = f_i(x)\}$$

for every $0 \le \alpha \le \bar{\alpha}$. Baire's Theorem states that at least one of the closed sets $\{x \in B(\bar{x}, \alpha) | f(x) = f_i(x)\}$, $i = 1, \ldots, k$, has an interior point, say x_α, which is also an interior point of the larger set $\{x \in O | f(x) = f_i(x)\}$ and thus an element of $O' \cap B(\bar{x}, \alpha)$. Hence if α_n is a null sequence of positive reals with $\alpha_n \le \bar{\alpha}$, then the sequence of vectors x_{α_n} is contained in O' and converges to $\bar{x}$.

So we can construct for each $x \in U$ a neighborhood O_x and an open subset $O'_x \subseteq O_x$ which is dense in O_x and has the property that f is r-times continuously differentiable at every point in O_x. Taking the union of all set O_x, we obtain an open set which is dense in U and has the property that f is r-times continuously differentiable at every point in the set. □

4.1.5 Comments and References

Piecewise differentiable functions are not coherently defined in the literature. The study of these functions started with Whitehead's paper [79] on C^1-complexes, where he introduced piecewise differentiable mappings on polyhedral subdivisions (cf. also [29]). In view of Proposition 4.1.4 such functions are strongly B-differentiable and hence the inverse function theorem of Sect. 3.2.2 is applicable. The notion of continuous selections of differentiable functions has been introduced in the article [27] of Jongen and Pallaschke to extend the classical critical point theory to nonsmooth functions (cf. also [3, 41]). Real-valued piecewise differentiable functions have been introduced by Womersley in [80] and extensively studied by Chaney in connection with nonsmooth optimization problems (cf. [7] and the references herein). For further results on real-valued PC^r-functions we refer to [38]. Our presentation of the subject is close to the exposition in [40], where also most of the results can be found. Continuous selections of locally Lipschitz continuous functions have been studied in Hager's paper [23]. For an account on Rademacher's Theorem we refer to Federer's book [17]. Baire's Theorem can be found in the book [32] of Kolmogorov and Fomin.

4.2 Piecewise Differentiable Homeomorphisms

In this section, we will be interested in the local homeomorphism property for piecewise differentiable functions. A function $f : U \to \mathbb{R}^n$ which is a local homeomorphism at $x_0 \in U \subseteq \mathbb{R}^n$ is called a *local PC^r-homeomorphism (local C^r-diffeomorphism)* at x_0 if f as well as its local inverse function are PC^r-functions (C^r-functions) in a neighborhood of x_0 and $f(x_0)$, respectively. The next result shows that we may apply any inverse function theorems for locally Lipschitz continuous functions to recognize PC^r-homeomorphisms.

Proposition 4.2.1. *A PC^r-function $f : U \to \mathbb{R}^n$ is a local PC^r-homeomorphism at $x_0 \in U \subseteq \mathbb{R}^n$ if and only if it is a local Lipschitz homeomorphism at x_0.*

Proof. Since by Corollary 4.1.1 a PC^r-function is locally Lipschitz continuous, every local PC^r-homeomorphism is certainly a local Lipschitz homeomorphism. To see the converse, let f be a local Lipschitz homeomorphism at x_0. Lemma 3.2.2 shows that there exists a neighborhood $O \subseteq U$ of x_0 and a constant $l > 0$ such that

$$\|f(x + y) - f(x)\| \geq l\|y\| \tag{4.9}$$

for every $x, y \in \mathbb{R}^n$ with $x, x + y \in O$. Moreover, since f is PC^r, it is B-differentiable at every point $x \in O$ and thus we can find for every $\varepsilon > 0$ a number $\delta > 0$ such that

$$\| f(x + y) - f(x) - f'(x; y) \| \leq \varepsilon \| y \| \qquad (4.10)$$

for every $\| y \| \leq \delta$. In view of the positive homogeneity of $f'(x; .)$, we deduce from (4.9) and (4.10) that

$$\| f'(x; y) \| \geq l \| y \| \qquad (4.11)$$

for every $y \in \mathbb{R}^n$. Reducing the neighborhood O if necessary, we may select a collection of C^r-functions $f_1, \dots, f_k : O \to \mathbb{R}^n$ which are essentially active selection functions for f at x_0. Hence for each $i \in \{1, \dots, k\}$ there exists a sequence of points $x_m^i \in U$, $m \in \mathbb{N}$, converging to x_0 such that

$$x_m^i \in \operatorname{int}\{x \in U \mid f(x) = f_i(x)\}.$$

In particular, f is F-differentiable at every point x_m^i and

$$f'(x_m^i; y) = \nabla f_i(x_m^i) y.$$

The inequality (4.11) thus yields

$$\| \nabla f_i(x_m^i) y \| \geq l \| y \|$$

for every $y \in \mathbb{R}^n$. Due to the continuity of ∇f_i, the latter inequality holds also for the limit point x_0. Hence $\nabla f_i(x_0) y = 0$ if and only if $y = 0$ and thus $\nabla f_i(x_0)$ is a nonsingular $n \times n$-matrix. The classical inverse function theorem for C^r-function thus shows that every selection function f_i is a local C^r-diffeomorphism at x_0. Since f was assumed to be a local Lipschitz homeomorphism, its local inverse is continuous. To prove that f^{-1} is a PC^r-function, it thus suffices to show that

$$f^{-1}(v) \in \{f_1^{-1}(v), \dots, f_k^{-1}(v)\} \qquad (4.12)$$

for every v close to $f(x_0)$, where the function f_i^{-1} is the local inverse function of f_i defined in a neighborhood of $f(x_0)$. Note that $f(x) = v$ for some $x \in O$ if and only if there exists a selection function $f_i(x) = v$. If v is close enough to $f(x_0)$, then the latter equation has a unique solution $x = f_i^{-1}(v)$ in a neighborhood of x_0, which proves (4.12) for every v in a sufficiently small neighborhood of $f(x_0)$.   □

The following inverse function theorem is the essence of this chapter.

Theorem 4.2.1. *Let $f : U \to \mathbb{R}^m$ be a PC^r-function and let $f_1, \dots, f_k : O \to \mathbb{R}^m$ be a collection of C^r-selection functions for f at $x_0 \in O \subseteq U$. If there exists a neighborhood $V \subseteq O$ of x_0 and a collection $C_1, \dots, C_l \subseteq \mathbb{R}^n$ of closed cones such that*

1. $\mathbb{R}^n = \cup_{i=1}^{l} C_i$
2. For every $i \in \{1, \ldots, l\}$ there exists an index $p(i) \in \{1, \ldots, k\}$ such that $f(x) = f_{p(i)}(x)$ for every $x \in V \cap (\{x_0\} + C_i)$

then f is a local PC^r-homeomorphism at x_0 if and only if its B-derivative $f'(x_0; .)$ is a homeomorphism.

Proof. Proposition 4.2.1 shows that f is a PC^r-homeomorphism at x_0 if and only if it is a local Lipschitz homeomorphism at x_0. In view of Proposition 4.1.4 the function f is strongly B-differentiable and thus Corollary 3.2.1 shows that f is a local Lipschitz homeomorphism at x_0 if and only if $f'(x_0; .)$ is a local Lipschitz homeomorphism at the origin. In view of Proposition 3.2.1, the positively homogeneous function $f'(x_0; .)$ is a local Lipschitz homeomorphism at the origin if and only if it is a Lipschitz homeomorphism. By Proposition 2.3.1 a piecewise affine homeomorphism has a piecewise affine inverse function and thus, in view of Proposition 2.2.7, a piecewise affine function is a homeomorphism if and only if it is a Lipschitz homeomorphism which completes the proof. □

4.2.1 An Implicit Function Theorem

Theorem 4.2.1 can be used in connection with Lemma 3.2.1 to prove the following implicit function theorem for PC^r-functions. Instead of formulating the result in its most general form, we use the theory of piecewise affine functions, in particular Theorem 2.3.7, to make the result accessible for many applications.

Theorem 4.2.2. *Let $U \subseteq \mathbb{R}^n \times \mathbb{R}^m$ be open, $f : U \to \mathbb{R}^n$ be a PC^r-function, and let*

$(x^0, y^0) \in U$ *be a vector with $f(x^0, y^0) = 0$,*
$f_1, \ldots, f_k : O \to \mathbb{R}^n$ *be a collection of selection functions for f at $(x^0, y^0) \in O \subseteq U$,*
Σ *be a conical subdivision of $\mathbb{R}^n \times \mathbb{R}^m$ with a lineality space of dimension l.*

If

1. *For every $\sigma \in \Sigma$ there exists an index $j_\sigma \in \{1, \ldots, k\}$ such that $f(x, y) = f_{j_\sigma}(x, y)$ for every $(x, y) \in O \cap (\sigma + \{(x^0, y^0)\})$*
2. *Either $(n + m - l) \leq 1$ or there exists a natural number $k \in \{2, \ldots, (n + m - l)\}$ such that the kth branching number of Σ does not exceed $2k$*
3. *All matrices $\nabla_x f_{j_\sigma}(x^0, y^0)$, $\sigma \in \Sigma$, have the same nonvanishing determinant sign*

then

1. *The equation $f(x, y) = 0$ determines an implicit PC^r-function $x(y)$ at (x^0, y^0)*
2. *The implicit functions $x_{j_\sigma}(y)$ determined by the equations $f_{j_\sigma}(x, y) = 0, \sigma \in \Sigma$, form a collection of selection functions for the PC^r-function $x(y)$ at y^0*

3. For every $w \in \mathbb{R}^m$ the identity $v = x'(y^0; w)$ holds if and only if v satisfies the piecewise linear equation $f'((x^0, y^0); (v, w)) = 0$

Proof. Having the classical approach for the construction of implicit function theorems from inverse function theorems in mind (cf. Lemma 3.2.1 and Remark 3.2.1), we first prove that the function $F : U \to \mathbb{R}^n \times \mathbb{R}^m$ defined by

$$F(x, y) = (f(x, y), y)$$

is a PC^r-homeomorphism at (x^0, y^0). The assumptions of the theorem ensure that the function F satisfies the assumptions of Theorem 4.2.1. Thus it suffices to prove that the B-derivative of F is a homeomorphism. Due to assumption 1 and the fact that the sets σ are cones, we deduce from the definition of the B-derivative that

$$F'((x^0, y^0); (v, w)) = (\nabla_x f_{j_\sigma}(x^0, y^0)v + \nabla_y f_{j_\sigma}(x^0, y^0)w, w) \tag{4.13}$$

for every $(v, w) \in \sigma$. Hence the conical subdivision Σ corresponds to the B-derivative of F at (x^0, y^0). In view of assumption 2 and Theorem 2.3.7, the B-derivative of F at (x^0, y^0) is a homeomorphism if and only if it is coherently oriented which, in view of (4.13), is equivalent to the fact that the matrices

$$\begin{pmatrix} \nabla_x f_{j_\sigma}(x^0, y^0) & \nabla_y f_{j_\sigma}(x^0, y^0) \\ 0 & I \end{pmatrix}, \sigma \in \Sigma, \tag{4.14}$$

have the same nonvanishing determinant sign. Performing successive Laplace expansions over the last m rows shows that the determinants of the latter matrices coincide with the determinants of the matrices $\nabla_x f_{j_\sigma}(x^0, y^0)$, which have the same nonvanishing sign by assumption. Hence F is a PC^r-homeomorphism at (x^0, y^0) and thus Lemma 3.2.1 and Remark 3.2.1 show that the equation $f(x, y) = 0$ determines an implicit function at (x^0, y^0) and that $x(y) = F_x^{-1}(0, y)$, which is locally a PC^r-function since F^{-1} is locally PC^r.

To prove the second assertion, we note that the functions $F_{j_\sigma}(x, y) = (f_{j_\sigma}(x, y), y)$ form a collection of selection functions for F at (x^0, y^0), and that all these selection functions are local C^r-diffeomorphisms at (x^0, y^0) since the corresponding Jacobians have nonvanishing determinant. Hence the locally defined functions $F_{j_\sigma}^{-1}$ form a collection of selection functions of the PC^r-function F^{-1} at $F(x^0, y^0)$ and thus $x(y) = F_x^{-1}(0, y)$ is a continuous selection of the C^r-functions $x_{j_\sigma}(y) = (F_{j_\sigma})_x^{-1}(0, y), \sigma \in \Sigma$, in a neighborhood of y^0.

In order to see the final assertion, let

$$\Phi(y) := (x(y), y) = F^{-1}(0, y). \tag{4.15}$$

Hence

$$\Phi'(y^0; w) = (x'(y^0; w), w)$$

$$= (F^{-1})'((0, y^0); (0, w)). \tag{4.16}$$

In view of part 1 of Theorem 3.2.3, $(F^{-1})'((0, y^0); (., .))$ is a homeomorphism and the identity

$$(v, w) = (F^{-1})'((0, y^0); (0, w))$$

holds if and only if

$$F'((x^0, y^0); (v, w)) = (0, w). \qquad (4.17)$$

In view of (4.16) and the definition of F, we thus obtain that $v = x'(y^0; w)$ if and only if v satisfies the equation

$$f'((x^0, y^0); (v, w)) = 0. \qquad \square$$

Beside the determination of $x'(x^0; w)$ by solving a piecewise linear equation, there is also a combinatorial result which may be used to determine the latter derivative:

Proposition 4.2.2. *Suppose the assumptions of Theorem 4.2.2 are satisfied and $w \in \mathbb{R}^m$.*

1. There exists a cone $\sigma \in \Sigma$ such that

$$\begin{pmatrix} 0 \\ w \end{pmatrix} \in \begin{pmatrix} \nabla_x f_{j_\sigma}(x^0, y^0) \ \nabla_y f_{j_\sigma}(x^0, y^0)w, w) \\ 0 \qquad\qquad I \end{pmatrix} \sigma \qquad (4.18)$$

2. The inclusion (4.18) holds if and only if

$$(-\nabla_x f_{j_\sigma}(x^0, y^0)^{-1} \nabla_y f_{j_\sigma}(x^0, y^0)w, w) \in \sigma \qquad (4.19)$$

3. If $w \in \mathbb{R}^m$ satisfies (4.18), then

$$x'(y^0; w) = -\nabla_x f_{j_\sigma}(x^0, y^0)^{-1} \nabla_y f_{j_\sigma}(x^0, y^0)w$$

Proof. We use the notations of the proof of Theorem 4.2.2.

1. To see the first assertion, note that every cone σ is mapped by F_{j_σ} onto a set $M_\sigma \in \mathbb{R}^n \times \mathbb{R}^m$ and that the sets M_σ, $\sigma \in \Sigma$, cover a neighborhood of $F(x^0, y^0) = (0, y^0)$ in the image space $\mathbb{R}^n \times \mathbb{R}^m$ since the function $F(x, y) = (f(x, y), y)$, which is a continuous selection of the functions F_{j_σ}, is a PC^r-homeomorphism at (x^0, y^0). Set $z^0 = (x^0, y^0)$ and fix a vector $\xi \in \mathbb{R}^n \times \mathbb{R}^m$. The surjectivity property yields the existence of a cone $\sigma \in \Sigma$, a null sequence α_k, and a sequence $z_k \in \sigma$ such that

$$F_{j_\sigma}(z^0) + \alpha_k \xi = F_{j_\sigma}(z^0 + z_k). \qquad (4.20)$$

Since F_{j_σ} is a local diffeomorphism at z^0, the sequence z_k can be chosen so that it converges to the origin. Passing to a subsequence if necessary, we may also assume that

$$\lim_{k\to\infty} \frac{z_k}{\|z_k\|} = \bar{z}.$$

Note that $\bar{z} \in \sigma$ since the latter cone is closed. The differentiability of F_{j_σ} yields

$$\lim_{k\to\infty} \frac{F_{j_\sigma}(z^0 + z_k) - F_{j_\sigma}(z^0)}{\|z_k\|} = \nabla F_{j_\sigma}(z^0)\bar{z}.$$

In view of (4.20) we thus obtain

$$\xi \lim_{k\to\infty} \frac{\alpha_k}{\|z_k\|} = \nabla F_{j_\sigma}(z^0)\bar{z},$$

which shows that $\xi \in \nabla F_{j_\sigma}(z^0)\sigma$.

2. The equivalence of the inclusions (4.18) and (4.19) is readily verified since the inverse of the matrix

$$\begin{pmatrix} \nabla_x f_{j_\sigma}(x^0, y^0) & \nabla_y f_{j_\sigma}(x^0, y^0) \\ 0 & I \end{pmatrix} \tag{4.21}$$

is given by

$$\begin{pmatrix} \nabla_x f_{j_\sigma}(x^0, y^0)^{-1} & -\nabla_x f_{j_\sigma}(x^0, y^0)^{-1}\nabla_y f_{j_\sigma}(x^0, y^0) \\ 0 & I \end{pmatrix}. \tag{4.22}$$

3. To see the second assertion of the proposition, recall from the last part of Theorem 4.2.2 that

$$x'(y^0; w) = v \text{ if and only if } f'((x^0, y^0); (v, w)) = 0. \tag{4.23}$$

Now let $w \in \mathbb{R}^m$ be a vector satisfying (4.19) and set

$$v = -\nabla_x f_{j_\sigma}(x^0, y^0)^{-1}\nabla_y f_{j_\sigma}(x^0, y^0)w \tag{4.24}$$

the inclusion (4.19) and assumption 1 of the theorem imply that

$$f(x^0 + \alpha v, y^0 + \alpha w) = f_{j_\sigma}(x^0 + \alpha v, y^0 + \alpha w)$$

for every sufficiently small $\alpha \geq 0$. Hence it follows immediately from the definition of the B-derivative that

$$f'((x^0, y^0); (v, w)) = f'_{j_\sigma}((x^0, y^0); (v, w)). \tag{4.25}$$

In view of (4.24), we obtain

$$f'_{j_\sigma}((x^0, y^0); (v, w))$$

$$= -\nabla_x f_{j_\sigma}(x^0, y^0)\nabla_x f_{j_\sigma}(x^0, y^0)^{-1}\nabla_y f_{j_\sigma}(x^0, y^0)w + \nabla_y f_{j_\sigma}(x^0, y^0)w = 0,$$

and thus (4.23) and (4.25) yield

$$x'(y^0; w) = -\nabla_x f_{j_\sigma}(x^0, y^0)^{-1}\nabla_y f_{j_\sigma}(x^0, y^0)w. \qquad \square$$

4.2.2 A Bound for the Condition Number

In the latter section we have developed an implicit function theorem for parametric problems which can be formulated as a parametric PC^1-function. The theory of condition is concerned with the behavior of the solution function to a parametric problem in the vicinity of a given parameter value. We will not go into the details of the theory of condition but instead only investigate the asymptotic condition number of the solution function. The *absolute asymptotic condition number* of a function $f : U \to \mathbb{R}^m$ defined on an open set $U \subseteq \mathbb{R}^n$ at a point $x^0 \in U$ is defined by

$$\kappa(f, x^0) = \limsup_{x \to x^0} \frac{\|f(x) - f(x^0)\|}{\|x - x^0\|}. \tag{4.26}$$

Since we are not dealing with other types of condition numbers, we will call $\kappa(f, x^0)$ simply the condition number of f at x^0. Note that $\kappa(f, x^0)$ provides some information about possible changes of the function values of f in terms of changes of the argument. In fact, for every $\varepsilon > 0$ there exists a $\delta > 0$ such that

$$\|f(x) - f(x^0)\| \le (\kappa(f, x^0) + \varepsilon)\|x - x^0\|$$

for every $\|x - x^0\| < \delta$. The following proposition shows how to calculate the condition number of a B-differentiable function.

Proposition 4.2.3. *If $f : U \to \mathbb{R}^m$ is a continuous B-differentiable function defined on an open set $U \subseteq \mathbb{R}^n$ and $x^0 \in U$, then*

$$\kappa(f, x^0) = \max_{\|y\|=1} f'(x^0; y).$$

Proof. Since f is B-differentiable, the continuity of f implies the continuity of the B-derivative $f'(x^0; .)$ Hence there exists a vector $y^* \in \mathbb{R}^n$ with $\|y^*\| = 1$ and

$$f'(x^0, y^*) \ge f'(x^0, y)$$

for every $y \in \mathbb{R}^n$ with $\|y\| = 1$. Thus we have to prove the identity

$$\limsup_{x \to x^0} \frac{\|f(x) - f(x^0)\|}{\|x - x^0\|} = \|f'(x^0, y^*)\|. \tag{4.27}$$

To see that the left-hand side does not exceed the right-hand side, recall that the definition of the B-derivative yields for every $\varepsilon > 0$ the existence of a number $\delta > 0$ such that

$$\|f(x) - f(x^0)\| \le \|f(x) - f(x^0) - f'(x^0; x - x^0)\| + \|f'(x^0; x - x^0)\|$$
$$\le \varepsilon\|x - x^0\| + \|f'(x^0; y^*)\|\|x - x^0\|$$

for every $x \in U$ with $\|x - x^0\| < \delta$. Hence

$$\limsup_{x \to x^0} \frac{\|f(x) - f(x^0)\|}{\|x - x^0\|} \le \varepsilon + \|f'(x^0, y^*)\|$$

for every $\varepsilon > 0$, which proves that the left-hand side of (4.27) is at most as large as the right-hand side. To see the converse inequality, note that by definition of the B-derivative

$$\lim_{\substack{t \to 0 \\ t > 0}} \frac{f(x^0 + ty^*) - f(x^0)}{t} = f^*(x^0; y^*).$$

Setting $x_n = x^0 + \frac{1}{n}y^*$ and recalling that $\|y^*\| = 1$, we thus obtain

$$\lim_{n \to \infty} \frac{\|f(x_n) - f(x^0)\|}{\|x_n - x^0\|} = \|f^*(x^0; y^*)\|$$

which shows that the left-hand side of (4.27) is at least as large as the right-hand side and thus completes the proof.   □

The following corollary is an immediate consequence of the latter proposition and the fact that the B-derivative is a continuous selection of the F-derivatives of a set of essentially active selection functions. Recall that the norm of an $m \times n$-matrix A is defined by

$$\|A\| = \max_{\|y\|=1} \|Ay\|.$$

Corollary 4.2.1. *If $U \subseteq \mathbb{R}^n$ is open, $f : U \to \mathbb{R}^m$ is a piecewise differentiable function, and $f_1, \ldots, f_k : O \to \mathbb{R}^m$ is a collection of selection functions for f at $x^0 \in O \subseteq U$, then*

$$\kappa(f, x^0) \le \max_{i \in I_f^e(x^0)} \|\nabla f_i(x^0)\|.$$

Proof. Since $f'(x^0; y) \in \{\nabla f_i(x^0) y \mid i \in I_f^e(x^0)\}$, one readily verifies that

$$\max_{\|y\|=1} \|f'(x^0; y)\| \leq \max_{\|y\|=1} \max_{i \in I_f^e(x^0)} \|\nabla f_i(x^0) y\|$$

$$= \max_{i \in I_f^e(x^0)} \max_{\|y\|=1} \|\nabla f_i(x^0) y\|$$

$$= \max_{i \in I_f^e(x^0)} \|\nabla f_i(x^0)\|. \qquad \square$$

4.2.3 Comments and References

A generalization of the inverse function Theorem 4.2.1 is given in [40], while a more general formulation of the implicit function Theorem 4.2.2 can be found in the paper [61] of Robinson. For a degree theoretic approach towards inverse and implicit function theorems for PC^r-functions we refer to Pang's article [52] and to the recent paper [54] of Pang and Ralph. The absolute asymptotic condition number has been introduced in Rice's paper [60] on the theory of condition which contains a nice introduction to the subject.

In [40, 54] more general conditions have been studied which relate local invertibility of a piecewise differentiable function to invertibility of its B-derivative. The paper [40] extends the differential topology approach to PC^1-maps initiated by Jongen and Pallaschke in their seminal paper [27] to vector-valued functions. One of the results in [40] states that a PC^1-map has a local PC^1-inverse at a point x if its B-derivative at x is invertible and every collection of at most n of the gradients of the essentially active selection functions of the components is linearly independent. The approach of Pang and Ralph in [54] is based on degree theory. They show, e.g., that a PC^1-function has a local PC^1-inverse at x if its B-derivative at x is invertible and every Jacobian of the essentially active selection function of the original function forms an essentially active selection function of the B-derivative. In [44, 58] the degree theoretic approach has been used to prove the following implicit function theorem:

Theorem 4.2.3. *Let $U \subseteq \mathbb{R}^n$, $V \subseteq \mathbb{R}^m$ be open sets and $f : U \times V \to \mathbb{R}^n$ be a PC^r-map with C^r-selection functions $f_1, \ldots, f_m$. Then the following statements are equivalent:*

1. *The equation $f(x, y) - z = 0$ determines an implicit PC^r-function $x(y, z)$ in a neighborhood of a solution (x_0, y_0, z_0).*
2. *The matrices $\nabla_x f_i(x_0, y_0)$, $i \in I_f^e(x_0, y_0)$, have the same nonvanishing determinant sign and the piecewise linear equation*

$$f'((x_0, y_0); (u, v)) - w = 0$$

has a unique solution $u(v, w)$ for every $v \in \mathbb{R}^m$ and $w \in \mathbb{R}^n$.

For $m = 0$, one obtains the most general PC^r-inverse function theorem. The latter theorem has been applied to mathematical programs with normal map constraints in [44] and to composite nonsmooth equations in [58].

4.3   Appendix: A Formula for the Generalized Jacobian

Generally the calculation of Clarke's generalized Jacobian can be quite difficult due to the lack of exact calculus rules. For piecewise differentiable functions, however, there is a representation of the generalized Jacobian at hand once a set of essentially active selection functions is known. A similar result is stated as Proposition 4 in [34].

Proposition 4.3.1. *If U is an open subset of $\mathbb{R}^n$ and $f : U \to \mathbb{R}^m$ is a PC^1-function with C^1 selection functions $f_i : O \to \mathbb{R}^m$, $i = 1, \ldots, k$, at $x_0 \in O \subseteq U$, then*

$$\partial f(x_0) = \mathrm{conv}\{\nabla f_i(x_0) | i \in I_f^e(x_0)\}.$$

Proof. By definition

$$I_f^e(x_0) = \{i \in \{1, \ldots, k\} | x_0 \in \mathrm{cl}(\mathrm{int}\{x \in O | f(x) = f_i(x)\})\};$$

hence for every $i \in I_f^e(x_0)$ there exists a sequence of points $x_n \in \mathrm{int}\{x \in O | f(x) = f_i(x)\}$, and thus f is Fréchet differentiable at x_n and $\nabla f(x_n) = \nabla f_i(x_n)$. The continuity of ∇f_i implies $\nabla f_i(x_0) \in \partial f(x_0)$, which proves the inclusion

$$\partial f(x_0) \supseteq \mathrm{conv}\{\nabla f_i(x_0) | i \in I_f^e(x_0)\}.$$

To see the converse inclusion, let x_n be a sequence of points converging to x_0 such that f is F-differentiable at each point x_n and that the sequence of Jacobians $\nabla f(x_n)$ converges. This implies that the B-derivative $f'(x_n; .)$ is a linear function. Recall from Proposition 4.1.1 that f is locally a continuous selection of the functions f_i, $i \in I_f^e(x_0)$ and from Proposition 4.1.3 that the B-derivative of a PC^1-function is a continuous selection of the F-derivatives of the selection functions. Hence for each sufficiently large $n \in \mathbb{N}$ there exists an index $i_n \in I_f^e(x_0)$ such that $\nabla f(x_n) = \nabla f_{i_n}(x_n)$. Considering the subsequences with constant indices, one readily verifies that

$$\lim_{n \to \infty} \nabla f(x_n) \in \{\nabla f_i(x_n) | i \in I_f^e(x_0)\}.$$

This proves the inclusion

$$\partial f(x_0) \subseteq \mathrm{conv}\{\nabla f_i(x_0) | i \in I_f^e(x_0)\},$$

and thus completes the proof of the theorem.   □

Chapter 5
Sample Applications

5.1 Variational Inequalities and Normal Maps

The notion of a normal map has been introduced by S.M. Robinson as a device for the treatment of nonlinear variational inequalities. Here, a *variational inequality* induced by a closed convex set $S \subseteq \mathbb{R}^n$ and a function $f : S \to \mathbb{R}^n$ is the problems of finding a vector $x \in S$ which satisfies

$$f(x)^T x \geq f(x)^T y \text{ for every } y \in S. \tag{5.1}$$

In view of the definition (2.2) of the normal cone, the variational inequality can be equivalently formulated as a so-called *generalized equation*

$$0 \in \{-f(x)\} + N_S(x). \tag{5.2}$$

Recall that by Proposition 2.4.2 the identity

$$N_S(x) + \{x\} = \Pi_S^{-1}(x) \tag{5.3}$$

holds, where Π_S is the Euclidean projection onto S. Hence the variational inequality (5.1) as well as the generalized equation (5.2) are equivalent to the fixed-point equation

$$\Pi_S(x + f(x)) = x. \tag{5.4}$$

The latter equation is closely related to the so-called *normal equation*

$$f(\Pi_S(z)) + \Pi_S(z) = z. \tag{5.5}$$

In fact, if x solves (5.4), then $z = x + f(x)$ solves (5.5), while $x = \Pi_S(z)$ solves (5.4) whenever z solves (5.5). The normal equation (5.5) determines the zeros of the mapping

$$f_S = f \circ \Pi_S + \Pi_S - I.$$

S. Scholtes, *Introduction to Piecewise Differentiable Equations*, SpringerBriefs in Optimization, DOI 10.1007/978-1-4614-4340-7_5, © Stefan Scholtes 2012

S.M. Robinson called the latter mapping the *normal map* induced by the function f and the closed convex set S. Naturally the question arises why the latter mapping should be preferred to the mapping $\Pi_S \circ (I + f) - I$, the zeros of which correspond directly to the solutions of the variational inequality (5.1) in view of its equivalence to (5.4). In fact, we have successfully used the latter function in Sect. 2.4.3 to analyze affine variational inequalities. However, in the nonlinear case the latter formulation has a severe drawback. In most applications of variational inequalities, the function f is a C^1-function. In this case the normal map f_S is differentiable at all points where Π_S is differentiable, while the set of points where the function $\Pi_S \circ (I + f) - I$ fails to be differentiable may have a more complicated structure. To get an idea of this phenomenon, consider the real-valued functions $g(x) = f(|x|)$ and $h(x) = |f(x)|$ of a single variable x. If f is differentiable, then g is differentiable at all points except perhaps at the origin, while h may be nondifferentiable at all points where f vanishes. The following proposition exploits the property that the nondifferentiability points of f_S coincide with the nondifferentiability points of Π_S.

Proposition 5.1.1. *If $P \subseteq \mathbb{R}^n$ is a polyhedron and $f : \mathbb{R}^n \to \mathbb{R}^n$ is a C^r-function, then f_P is a strongly B-differentiable PC^r-function.*

Proof. Let $P = \{x \in \mathbb{R}^n | Ax \le b\}$ and suppose $a_1, \ldots, a_m$ are the row vectors of A. In view of Proposition 2.4.4 the function f_P is a continuous selection of the functions

$$f \circ \Pi_{S_I} + \Pi_{S_I} - I, I \in \mathscr{I}(A, b), \tag{5.6}$$

where $S_I = \{x \in \mathbb{R}^n | a_i^T x = b_i, i \in I\}$, and $\mathscr{I}(A, b)$ is defined by (2.42). By Proposition 2.4.3 the functions Π_{S_I} are affine. Hence the latter selection functions are C^r-functions and thus f_P is a PC^r-function. Moreover, Proposition 2.4.4 shows that f_P coincides with a selection function on each polyhedron of the normal manifold included by P. Since a polyhedral subdivision Σ of $\mathbb{R}^n$ coincides in a neighborhood of a point x with the polyhedral subdivision $\{\{x\} + \sigma' | \sigma' \in \Sigma'(x)\}$, where $\Sigma'(x)$ is the localization (2.23) of Σ at x, we deduce from Proposition 4.1.4 that f_P is indeed strongly B-differentiable at each point $x \in \mathbb{R}^n$. $\qquad\square$

5.1.1 A Homeomorphism Condition for Normal Maps of Polyhedra

As a consequence of Proposition 5.1.1 the normal map f_P is a PC^1-function, provided that $f : \mathbb{R}^n \to \mathbb{R}^n$ is a C^1-function and that P is a polyhedron. Proposition 4.1.5 thus yields the existence of an open and dense subset Ω of $\mathbb{R}^n$ such that f is continuously differentiable at every point $x \in \Omega$. The following theorem provides a global homeomorphism condition for the normal map f_P in terms of the determinants of the Jacobian of f_P in Ω.

Theorem 5.1.1. *Let $P \subseteq \mathbb{R}^n$ be a polyhedron, $f : \mathbb{R}^n \to \mathbb{R}^n$ be a C^r-function, and let Ω be a dense subset of $\mathbb{R}^n$ such that f_P is F-differentiable for every $x \in \Omega$.*

If all matrices $\nabla f_P(x)$, $x \in \Omega$, *have the same nonvanishing determinant sign and if there exists a positive real l such that* $\|\nabla f_P(x)^{-1}\| \leq l$ *for every* $x \in \Omega$, *then* f_P *is a global PC^r-homeomorphism.*

Proof. We prove the theorem for the case that the determinants of the matrices $\nabla f_P(x)$, $x \in \Omega$, are positive; the proof for the case of negative determinants is mutatis mutandis the same. If all determinants are positive, then the second assumption of the theorem implies that there exists a number $\gamma > 0$ such that

$$\det(\nabla f_P(x)) \geq \gamma \quad \text{for every } x \in \Omega. \tag{5.7}$$

As in the proof of the latter proposition we assume that $P = \{x \in \mathbb{R}^n | Ax \leq b\}$, where A has row vectors $a_1, \ldots, a_m$ and use Proposition 2.4.4 to deduce that f_P is a continuous selection of the C^r-functions

$$f_I = f \circ \Pi_{S_I} + \Pi_{S_I} - I, \, I \in \mathscr{I}(A, b), \tag{5.8}$$

where $S_I = \{x \in \mathbb{R}^n | a_i^T x = b_i, i \in I\}$, and that f_P coincides with f_I on the polyhedron $P_I = F_I + N_I$ of the normal manifold, i.e., given a point $x \in \mathbb{R}^n$, the function f_P is a continuous selection of the C^r-functions f_I, $x \in P_I$, in a neighborhood of x. Since the polyhedron P_I has nonempty interior, the set $\Omega \cap \text{int} P_I$ is dense in P_I and hence we deduce from (5.7) and the second assumption of the theorem that

$$\det(\nabla f_I(x)) \geq \gamma$$
$$\|\nabla f_I(x)^{-1}\| \leq l \tag{5.9}$$

Since the B-derivative $f_P'(x; .)$ is a continuous selection of the F-derivatives $\nabla f_I(x)$, $x \in P_I$, we deduce from the first inequality of (5.9) that every B-derivative $f_P'(x; .)$ is coherently oriented. Moreover, a conical subdivision of $\mathbb{R}^n$ corresponding to the piecewise linear function $f_P'(x; .)$ is the localization (2.23) $\Sigma'(x)$ of the normal manifold Σ induced by P. In view of Lemma 2.3.4, the second branching number of the localization does not exceed the second branching number of the normal manifold which, in view of Proposition 2.4.5, does not exceed 4. Thus Theorem 2.3.7 shows that $f_P'(x; .)$ is a homeomorphism. Since by Proposition 5.1.1 the function f_P is strongly B-differentiable, we thus conclude that f_P is a local PC^r-homeomorphism. Using the second inequality of (5.9), we obtain

$$\|f'(x; y)\| \geq \min_{I:x \in P_I} \|\nabla f_I(x)y\|$$

$$\geq \min_{I:x \in P_I} \frac{1}{\|\nabla f_I(x)^{-1}\|} \|y\|$$

$$\geq \frac{1}{l} \|y\|$$

for every $y \in \mathbb{R}^n$. Thus Hadamard's Theorem 3.2.5 shows that f_P is indeed a homeomorphism. $\qquad\square$

5.1.2 Comments and References

For an account on variational inequalities we refer to the recent survey article [25] of Harker and Pang. An introduction to generalized equations can be found in Robinson's paper [62]. Normal maps have been studied by Robinson in [61,63,66], by Pang and Ralph in [54], and by Kuntz and the author in [40]. The result presented here is complementary to the results in [63].

5.2 Sensitivity Analysis for Mathematical Programs

In this section we use the implicit function theorem to develop conditions which ensure that a parametric mathematical programming problem

$$\mathbf{P(y)} \quad \min_{x \in \mathbb{R}^n} \{ f(x, y) | g(x, y) \le 0, h(x, y) = 0 \}$$

has locally a unique solution and to investigate properties of the solution function. We assume throughout this section that the functions $f : \mathbb{R}^n \times \mathbb{R}^p \to \mathbb{R}$, $g : \mathbb{R}^n \times \mathbb{R}^p \to \mathbb{R}^l$, and $h : \mathbb{R}^n \times \mathbb{R}^p \to \mathbb{R}^m$ are C^r-functions with $r \ge 2$. The minimization is carried out over the variable $x \in \mathbb{R}^n$, while the vector $y \in \mathbb{R}^p$ serves as a parameter which might reflect uncertainties in the problem data or control variables. Throughout this section, we assume that we are given a stationary point x^0 of the mathematical program $P(y^0)$ for a fixed parameter vector $y^0 \in \mathbb{R}^p$ and we are interested in the following questions:

1. Is the solution x^0 locally unique for the fixed parameter vector y^0?
2. If the first question is answered affirmatively, does there exist a locally unique stationary point $x(y)$ close to x^0 if y is close to y^0?
3. If the second question is answered affirmatively, which conditions ensure that the stationary points $x(y)$ are local solutions of the programs $P(y)$ for y close to y_0?
4. If the second question is answered affirmatively, what are the properties of the stationary point mapping $x(y)$?

   (a) Is the function $x(y)$ piecewise differentiable?
   (b) Can we calculate a collection of selection functions at y^0?
   (c) Can we calculate the B-derivative of $x(y)$ at the point y^0?

We emphasize the necessity to find algorithmically verifiable conditions to treat the latter questions.

5.2.1 Sensitivity Analysis of Stationary Solutions

We have already seen in the introductory chapter that the stationary points of the mathematical programming problem $P(y)$ correspond to the zeros of the Kojima mapping $F : \mathbb{R}^n \times \mathbb{R}^{l+m} \times \mathbb{R}^p \to \mathbb{R}^n \times \mathbb{R}^{l+m}$ which is defined by

$$
F(x, v, y) := \begin{pmatrix} \nabla_x f(x, y) + \displaystyle\sum_{i=1}^{l} \max\{v_i, 0\} \nabla_x g_i(x, y) + \displaystyle\sum_{j=l+1}^{l+m} v_j \nabla_x h_j(x, y) \\ -g_1(x, y) + \min\{v_1, 0\} \\ \vdots \\ -g_l(x, y) + \min\{v_l, 0\} \\ -h(x, y) \end{pmatrix},
$$

i.e., x is a stationary point of the program $P(y)$ if and only if there exists a vector $v \in \mathbb{R}^{l+m}$ such that $F(x, v, y) = 0$. Clearly F is a PC^{r-1}-function, provided the data functions determining the program $P(y)$ are C^r-functions. The function F is a C^{r-1}-function in a neighborhood of any point (x^0, v^0, y^0) with nonvanishing components v_i^0, $i = 1, \ldots, l$. Note that in this case the stationary point x^0 admits Lagrange multipliers $(\lambda^0, \mu^0) \in \mathbb{R}^l \times \mathbb{R}^m$ which satisfy the strict complementarity condition, i.e., $\lambda_i > 0$ for every i with $g_i(x^0, y^0) = 0$. If some of the components v_i^0, $i \in \{1, \ldots, l\}$, are vanishing, then the function F is locally a continuous selection of C^{r-1}-functions. The objective of this section is to analyze the local properties of the function F in a neighborhood of a general point $(x^0, v^0, y^0) \in \mathbb{R}^n \times \mathbb{R}^{l+m} \times \mathbb{R}^p$. As a consequence of Theorem 4.2.2 we obtain the following sensitivity result for the stationary points of parametric programs.

Theorem 5.2.1. *Let* $(x^0, v^0, y^0) \in \mathbb{R}^n \times \mathbb{R}^{l+m} \times \mathbb{R}^p$ *be a zero of the Kojima mapping* F *corresponding to a parametric programming problem* $P(y)$ *with* C^r- *data,* $r \geq 2$. *Let*

$$
\Delta(v^0) = \{\delta \in \{-1, 1\}^l \mid \delta_i = \operatorname{sign} v_i^0 \text{ if } v_i^0 \neq 0\}, \tag{5.10}
$$

and define for $\delta \in \Delta(v^0)$ *the* C^{r-1}-*function*

$$
F_\delta(x, v, y) := \begin{pmatrix} \nabla_x f(x, y) + \displaystyle\sum_{i=1}^{l} \max\{\delta_i, 0\} v_i \nabla_x g_i(x, y) + \displaystyle\sum_{j=l+1}^{l+m} v_j \nabla_x h_j(x, y) \\ -g_1(x, y) - \min\{\delta_1, 0\} v_1 \\ \vdots \\ -g_l(x, y) - \min\{\delta_l, 0\} v_l \\ -h(x, y) \end{pmatrix}.
$$

If all matrices $\nabla_{(x,v)}F_\delta(x^0, v^0, y^0)$, $\delta \in \Delta(v^0)$, have the same nonvanishing determinant sign, then the following statements hold:

Implicit function: *The equation $F(x, v, y) = 0$ determines implicit PC^{r-1}-functions $x(y)$ and $v(y)$ at (x^0, v^0, y^0).*

Selection functions: *The collections of all implicit functions $x_\delta(y)$ and $v_\delta(y)$ determined by the equations $F_\delta(x, v, y) = 0, \delta \in \Delta(v^0)$, are collections of selection functions for $x(y)$ and $v(y)$.*

B-derivative of the solution function: *If we define*

$$A(x^0, v^0, y^0) = \nabla_{xx}^2 f(x^0, y^0) + \sum_{i=1}^{l} \max\{v_i^0, 0\}\nabla_{xx}^2 g_i(x^0, y^0)$$

$$+ \sum_{j=l+1}^{l+m} v_j^0 \nabla_{xx}^2 h_j(x^0, y^0),$$

$$B(x^0, v^0, y^0) = \nabla_{xy}^2 f(x^0, y^0) + \sum_{i=1}^{l} \max\{v_i^0, 0\}\nabla_{xy}^2 g_i(x^0, y^0)$$

$$+ \sum_{j=l+1}^{l+m} v_j^0 \nabla_{xy}^2 h_j(x^0, y^0),$$

$$I_0(v^0) = \left\{i \in \{1, \ldots, l\} | v_i^0 = 0\right\},$$

$$I_+(v^0) = \left\{j \in \{1, \ldots, l\} | v_j^0 > 0\right\},$$

$$I_-(v^0) = \left\{s \in \{1, \ldots, l\} | v_s^0 < 0\right\},$$

then for every $w \in \mathbb{R}^m$ the quadratic program

$$\textbf{QP}(\mathbf{x^0, v^0, y^0, w}) \min \frac{1}{2} z^T A\left(x^0, v^0, y^0\right) z + z^T B\left(x^0, v^0, y^0\right) w$$

subject to

$$\nabla_x g_i\left(x^0, y^0\right)^T z \leq -\nabla_y g_i\left(x^0, y^0\right)^T w, i \in I_0\left(v^0\right),$$

$$\nabla_x g_j\left(x^0, y^0\right)^T z = -\nabla_y g_j\left(x^0, y^0\right)^T w, j \in I_+\left(v^0\right),$$

$$\nabla_x h_k\left(x^0, y^0\right)^T z = -\nabla_y h_k\left(x^0, y^0\right)^T w, k = 1, \ldots, m,$$

has a unique stationary point $z(w)$ with unique corresponding Lagrange multipliers $(\lambda(w), \gamma(w), \mu(w)) \in \mathbb{R}_+^{I_0(v^0)} \times \mathbb{R}^{I_+(v^0)} \times \mathbb{R}^m$, and the relations

$$x'(y^0; w) = z(w),$$

$$v_i'(y^0; w) = \begin{cases} \lambda_i(w) & \text{if } i \in I_0(v^0) \text{ and } \lambda_i(w) > 0, \\ \nabla_x g_i(x^0, y^0)^T v(w) - \nabla_y g_i(x^0, y^0)w & \text{if } i \in I_0(v^0) \text{ and } \lambda_i(w) = 0 \\ & \text{or } i \in I_-(v^0), \\ \gamma_i(w) & \text{if } i \in I_+(v^0), \\ \mu_{i-l}(w) & \text{if } i \in \{l+1, \ldots, l+m\} \end{cases}$$

hold.

Proof. Since the nonvanishing sign of a number is locally constant, the set

$$O(v^0) = \{(x, v, y) \in \mathbb{R}^n \times \mathbb{R}^{l+m} \times \mathbb{R}^p | \text{sign} v_i = \text{sign} v_i^0 \text{ if } v_i^0 \neq 0\}$$

is an open neighborhood of (x^0, v^0, y^0). Setting

$$\sigma_\delta(v^0) = \{(v, \mu, w) \in \mathbb{R}^n \times \mathbb{R}^{l+m} \times \mathbb{R}^p | \delta_i \mu_i \geq 0, i \in I_0(v^0)\}, \tag{5.11}$$

the definitions of F_δ and $\Delta(v^0)$ show that for every $\delta \in \Delta(v^0)$ and every

$$(x, v, y) \in O(v^0) \cap (\sigma_\delta(v^0) + \{(x^0, v^0, y^0)\})$$

the identity $F(x, v, y) = F_\delta(x, v, y)$ holds. It is easily seen that the collection of all cones $\sigma_\delta(v^0)$, $\delta \in \Delta(v^0)$, is a conical subdivision of $\mathbb{R}^n \times \mathbb{R}^{l+m} \times \mathbb{R}^p$ with lineality space

$$L = \{(v, \mu, w) \in \mathbb{R}^n \times \mathbb{R}^{l+m} \times \mathbb{R}^p | \mu_i = 0, i \in I_0(v^0)\}.$$

Note that the codimension $(n + l + m + p - \dim L)$ of the lineality space L equals the cardinality of $I_0(v^0)$. If the codimension is larger than 1, then every face of codimension 2 is contained in exactly four cones. In fact, a face of codimension 2 of a cone is obtained by turning two inequalities into equalities. Thus this face is contained in the four cones which correspond to the four different possibilities to turn two equalities into inequalities. Hence the second branching number of the conical subdivision is 4. Since by assumption the matrices $\nabla_{(x,v)} F_\delta(x^0, v^0, y^0)$ have the same nonvanishing determinant sign, we may apply Theorem 4.2.2. The first two assertions follow immediately from the statements (1) and (2) of Theorem 4.2.2. To see that the third assertion is a direct consequence of Part 3 of Theorem 4.2.2, note that

$$F'((x^0, v^0, y^0); (v, \eta, w)) = 0$$

if and only if

$$A(x^0, v^0, y^0)v + B(x^0, y^0, y^0)w + \sum_{i \in I_0(v^0)} \max\{\eta_i, 0\} \nabla_x g_i(x^0, y^0)$$

$$+ \sum_{j \in I_+(v^0)} \eta_j \nabla_x g_j(x^0, y^0) + \sum_{k=1}^{m} \eta_{l+k} \nabla_x h_k(x^0, y^0) = 0,$$

$$- \nabla_x g_i(x^0, y^0)^T v - \nabla_y g_i(x^0, y^0)^T w + \min\{\eta_i, 0\} = 0, i \in I_0(v^0),$$

$$- \nabla_x g_j(x^0, y^0)^T v - \nabla_y g_j(x^0, y^0)^T w = 0, j \in I_+(v^0),$$

$$- \nabla_x g_s(x^0, y^0)^T v - \nabla_y g_s(x^0, y^0)^T w + \eta_s = 0, s \in I_-(v^0),$$

$$- \nabla_x h_k(x^0, y^0)^T v - \nabla_y h_k(x^0, y^0)^T w = 0, k = 1, \ldots, m.$$

Eliminating the forth set equations which has free η_s variables, the left-hand side of the latter equation coincides with the Kojima mapping corresponding to the quadratic program $QP(x^0, v^0, y^0, w)$. Thus the third assertion of the theorem follows immediately from statement (3) of Theorem 4.2.2.   □

In view of the definition of the functions F_δ and the set $\Delta(v^0)$, the reduced Jacobians of $F_\delta, \delta \in \Delta(v^0)$, have the following block structure:

$$\nabla_{(x,v)} F_\delta(x^0, v^0, y^0) = \begin{pmatrix} A & B_\delta & C \\ -D^T & E_\delta & 0 \\ -C^T & 0 & 0 \end{pmatrix}, \tag{5.12}$$

where

1. A is an $n \times n$-matrix with

$$A = \nabla_{xx}^2 f(x^0, y^0) + \sum_{i=1}^{l} \max\{\delta_i, 0\} v_i^0 \nabla_{xx}^2 g(x^0, y^0) + \sum_{j=l+1}^{m+1} v_j^0 \nabla_{xx}^2 h(x^0, y^0)$$

$$= \nabla_{xx}^2 f(x^0, y^0) + \sum_{i=1}^{l} \max\{v_i^0, 0\} \nabla_{xx}^2 g(x^0, y^0) + \sum_{j=l+1}^{m+1} v_j^0 \nabla_{xx}^2 h(x^0, y^0),$$

where the equality is due to the fact that $\delta \in \Delta(v^0)$ (cf. 5.10)
2. B_δ is an $n \times l$-matrix defined by

$$B_\delta = \left(\max\{\delta_1, 0\} \nabla_x g_1(x^0, y^0), \ldots, \max\{\delta_l, 0\} \nabla_x g_l(x^0, y^0) \right)$$

3. C is the $n \times m$-matrix $\nabla_x h(x^0, y^0)^T$
4. D is the $l \times n$-matrix $\nabla_x g(x^0, y^0)^T$
5. E_δ is an $l \times l$-matrix the ith column of which is the vector $-\min\{\delta_i, 0\} e_i$, where e_i is the ith unit vector in $\mathbb{R}^l$

A natural question is how we can exploit the block structure of the matrix (5.12) to simplify the calculation of the determinant sign of the matrices $\nabla_{(x,v)} F_\delta(x^0, v^0, y^0)$, $\delta \in \Delta(v^0)$. A first observation is the fact that for $\delta_i = -1$, the $(n + i)$th column of the matrix (5.12) is the $(n + i)$th unit vector in $\mathbb{R}^{n+l+m}$. Using the Laplace expansion for the determinant, we can thus delete the $(n + i)$th row and column of the matrix without changing the determinant. Definition 5.10 shows that for every $i \in \{1, \ldots, l\}$ with $v_i^0 < 0$ and every $\delta \in \Delta(v^0)$ the identity $\delta_i = -1$ holds. Deleting all columns and rows corresponding to negative components of δ, we thus conclude

Remark 5.2.1. For every $\delta \in \Delta(v^0)$ the identity

$$\det \left(\nabla_{(x,v)} F_\delta(x^0, v^0, y^0) \right) = \det \begin{pmatrix} A & M_\delta & C \\ -M_\delta^T & 0 & 0 \\ -C^T & 0 & 0 \end{pmatrix}, \qquad (5.13)$$

holds, where M_δ is a matrix the columns of which store the gradients $\nabla_x g_i(x^0, y^0)$ with $\delta_i = 1$.

Note that the columns of the matrix M_δ consist of all gradients of active inequality constraints with positive multipliers and, in addition, some of the gradients of active inequality constraints with vanishing multiplier, according to our choice of $\delta \in \Delta(v^0)$. Each such matrix is a submatrix of the largest of these matrices which consists of all column vectors $\nabla_x g_i(x^0, y^0)$ corresponding to active inequality constraints. If we denote this largest matrix by M, we thus obtain as a necessary condition for all matrices on the right-hand side of (5.13) to have a nonvanishing determinant that the block matrix (M, C) has full column rank. This condition is known as the *linear independence constraint qualification*. It implies that the Lagrange multiplier vector λ^0 corresponding to the stationary solution x^0 is unique, which in turn implies the uniqueness of v^0. We thus conclude

Remark 5.2.2. If all matrices $\nabla_{(x,v)} F_\delta(x^0, v^0, y^0)$, $\delta \in \Delta(v^0)$, have a nonvanishing determinant sign, then the program $P(y^0)$ satisfies the linear independence constraint qualification at the stationary point x^0.

Te latter fact enables us to use the following elementary lemma from linear algebra for a further exploitation of the block structure of the matrix on the right-hand side of (5.13).

Lemma 5.2.1. *Let A and B be $n \times n$-matrices and let $B = (V, W)$, where V is an $n \times (n - m)$-matrix and W is an $n \times m$-matrix. If $\det(B) \neq 0$ and $V^T W = 0$, then*

$$\det \begin{pmatrix} A & W \\ -W^T & 0 \end{pmatrix} = \frac{\det(W^T W)^2}{\det(B)^2} \det(V^T A V).$$

Proof. Consider the $(n + m) \times ((n - m) + m + m)$ block matrix

$$\begin{pmatrix} V & 0 & W \\ 0 & I & 0 \end{pmatrix}$$

Interchanging m columns yields

$$\det \begin{pmatrix} V & 0 & W \\ 0 & I & 0 \end{pmatrix} = (-1)^m \det \begin{pmatrix} B & 0 \\ 0 & I \end{pmatrix}$$

$$= (-1)^m \det(B)$$

$$\neq 0. \tag{5.14}$$

Since $V^T W = 0$, we obtain

$$\begin{pmatrix} V & 0 & W \\ 0 & I & 0 \end{pmatrix}^T \begin{pmatrix} A & W \\ -W^T & 0 \end{pmatrix} \begin{pmatrix} V & 0 & W \\ 0 & I & 0 \end{pmatrix} = \begin{pmatrix} V^T A V & 0 & V^T A W \\ 0 & 0 & -W^T W \\ W^T A V & W^T W & W^T A W \end{pmatrix}$$

and thus in view of (5.14)

$$\det \begin{pmatrix} A & W \\ -W^T & 0 \end{pmatrix} = \frac{1}{\det(B)^2} \det \begin{pmatrix} V^T A V & 0 & V^T A W \\ 0 & 0 & -W^T W \\ W^T A V & W^T W & W^T A W \end{pmatrix}. \tag{5.15}$$

Since the matrix B is nonsingular, the submatrix W has full column rank. Hence the matrix $W^T W$ is nonsingular and we can calculate

$$\begin{pmatrix} 0 & -W^T W \\ W^T W & W^T A W \end{pmatrix}^{-1} = \begin{pmatrix} (W^T W)^{-1} W^T A W (W^T W)^{-1} & (W^T W)^{-1} \\ -(W^T W)^{-1} & 0 \end{pmatrix}. \tag{5.16}$$

The latter identity shows that

$$(0, V^T A W) \begin{pmatrix} 0 & -W^T W \\ W^T W & W^T A W \end{pmatrix}^{-1} \begin{pmatrix} 0 \\ W^T A V \end{pmatrix} = 0$$

and thus the Schur complement formula yields

$$\det \begin{pmatrix} V^T A V & 0 & V^T A W \\ 0 & 0 & -W^T W \\ W^T A V & W^T W & W^T A W \end{pmatrix} = \det \begin{pmatrix} 0 & -W^T W \\ W^T W & W^T A W \end{pmatrix} \det(V^T A V)$$

$$= (-1)^m \det \begin{pmatrix} -W^T W & 0 \\ W^T A W & W^T W \end{pmatrix} \det(V^T A V)$$

$$= (-1)^m \det(-W^T W) \det(W^T W) \det(V^T A V)$$

$$= \det(W^T W)^2 \det(V^T A V).$$

Equation (5.15) proves the assertion of the lemma.   □

Setting $W = (M_\delta, C)$, where M_δ is the matrix defined in Remark 5.2.1, we can immediately apply the latter lemma to prove the following result:

Remark 5.2.3. If the program $P(y^0)$ satisfies the linear independence constraint qualification at x^0, then for every $\delta \in \Delta(v^0)$ the equality

$$\text{sign det} \begin{pmatrix} A & M_\delta & C \\ -M_\delta^T & 0 & 0 \\ -C^T & 0 & 0 \end{pmatrix} = \text{sign det}(V_\delta^T A V_\delta) \tag{5.17}$$

holds, where the columns of the matrix V_δ form a basis of the nullspace of the matrix $(M_\delta, C)^T$.

5.2.2 Sensitivity Analysis of Local Minimizers

In the preceding section, we have investigated the local change of a stationary point as a function of the parameter. Here, we will be more interested in the change of a local minimizer as a function of the parameter. We will see that the key towards the analysis of the local minimizers is played by the following second-order sufficiency condition.

Second-order sufficiency condition: If $x^0 \in \mathbb{R}^n$ is a stationary solution of the program $P(y^0)$ and $\lambda^0 \in \mathbb{R}^{l+m}$ is a corresponding Lagrange multiplier, then x^0 is a local minimizer provided that $w^T A w > 0$ for every nonvanishing vector $w \in \mathbb{R}^n$ satisfying

$$\nabla h(x^0, y^0) w = 0$$
$$\nabla g_i(x^0, y^0)^T w = 0 \text{ if } \lambda_i^0 > 0,$$
$$\nabla g_j(x^0, y^0)^T w \leq 0 \text{ if } \lambda_j^0 = 0 \text{ and } g_j(x^0, y^0) = 0,$$

where

$$A = \nabla_{xx}^2 f(x^0, y^0) + \sum_{i=1}^{l} \lambda_i^0 \nabla_{xx}^2 g(x^0, y^0) + \sum_{j=l+1}^{m+1} \lambda_j^0 \nabla_{xx}^2 h(x^0, y^0).$$

We have not incidentally denoted the latter matrix by the symbol A. It does indeed coincide with the matrix A used in the representation (5.12) of $\nabla_{(x,v)} F_\delta(x^0, v^0, y^0)$, where the vector

$$v_i^0 = \begin{cases} \lambda_i^0 & \text{if either } \lambda_i^0 > 0 \text{ or } i \in \{l+1,\ldots,l+m\}, \\ g_i(x^0, y^0) & \text{otherwise,} \end{cases} \tag{5.18}$$

is chosen in such a way that $F(x^0, v^0, y^0) = 0$.

Proposition 5.2.1. *Let x^0 be a stationary point of the mathematical program $P(y^0)$ and suppose the linear independence constraint qualification holds at x^0. Let $(\lambda^0, \mu^0) \in \mathbb{R}_+^l \times \mathbb{R}^m$ be the Lagrange multiplier vector corresponding to x^0, and let V be a matrix whose columns form a basis of the linear subspace*

$$L = \{z \in \mathbb{R}^n | \nabla_x h(x^0, y^0)z = 0, \nabla_x g_i(x^0, y^0)^T z = 0, \text{if } \lambda_i^0 > 0\}.$$

If the matrix $V^T A V$ is positive definite, then there exist open neighborhoods $U \subseteq \mathbb{R}^p$ of y^0 and $V \subseteq \mathbb{R}^n$ of x^0 such that for every parameter value $y \in U$ there exists a unique stationary solution $x(y) \in V$ of the program $P(y)$. Moreover, x is a PC^{r-1}-function of the parameter y and every stationary solution $x(y)$ is a local minimizer of the corresponding program.

Proof. Defining v^0 by (5.18), we obtain $F(x^0, v^0, y^0) = 0$. Note that $V^T A V$ is positive definite if and only if $w^T A w > 0$ for every nonvanishing vector $w \in L$. If we define

$$L_\delta = \left\{ z \in \mathbb{R}^n | \nabla_x h(x^0, y^0)z = 0, \nabla_x g_i(x^0, y^0)^T z = 0, \text{if } v_i^0 > 0 \right.$$

$$\left. \nabla_x g_j(x^0, y^0)^T z = 0, \text{if } v_i^0 = 0 \text{ and } \delta_i = 1 \right\},$$

then $L_\delta \subseteq L$ and thus $w^T A w > 0$ for every nonvanishing vector $w \in L_\delta$. If the columns of V_δ form a basis of the subspace L_δ, we can thus deduce that the matrix $V_\delta^T A V_\delta$ is positive definite and as thus has a positive determinant. In view of the linear independence assumption, Remarks 5.2.1 and 5.2.3 and Theorem 5.2.1 can be readily used to prove the local existence of a piecewise differentiable stationary point mapping $x(y)$. To see that for y close enough to y^0 the stationary point $x(y)$ is a local minimizer, note that the assumptions of the corollary imply the second-order sufficiency condition. Since x depends continuously on the parameter y, so do the matrix A as well as the equality system defining L. Since $w^T A w > 0$ for every nonvanishing vector $w \in L$ if and only if the latter inequality holds for every $w \in L$ with unit norm, we deduce that $w^T A(y)w > 0$ for every $w \in L(y)$ as long as y is sufficiently close to y^0. $\qquad\square$

Remark 5.2.4. If the assumptions of the corollary are satisfied, then the objective function of the corresponding quadratic program $QP(x^0, v^0, y^0, w)$ exhibited in Theorem 5.2.1 is strictly convex over the feasible region and thus its unique stationary solution is a minimizer.

5.2.3   Comments and References

The standard introductory text to sensitivity analysis is Fiacco's book [18]. Our treatment of the subject follows the approach of Kojima in [31]. He called a stationary solution which satisfies the conditions of Theorem 5.2.1 a *strongly stable stationary solution.* A survey of equivalent characterizations can be found in the paper [28] of Klatte and Tammer. The directional derivative of the solution function has been investigated under weaker assumptions in a number of articles (cf. e.g. [2, 4, 5, 13, 21, 42, 57, 76]).

References

1. Alexandroff, P., Hopf, H.: Topologie. Chelsea Publishing Company, New York (1972)
2. Auslender, A., Cominetti, R.: First and second order sensitivity analysis of nonlinear programs under directional constraint qualification condition. Optimization **21**, 351–363 (1990)
3. Bartels, SG., Kuntz, L., Scholtes, S.: Continuous selections of linear functions and nonsmooth critical point theory. Nonlinear Anal. Theor. Meth. Appl. **24**, 385–407 (1994)
4. Bonnans, JF.: Directional derivatives of optimal solutions in smooth nonlinear programming. J. Optim. Theor. Appl. **73**, 27–45 (1992)
5. Bonnans, JF., Ioffe, AD., Shapiro, A.: Expansion of exact and approximate solutions in nonlinear programming. In: Oettli, W., Pallaschke, D. (eds.) Advances in Optimization. Springer Verlag, Berlin (1992)
6. Browder, FE.: Covering spaces, fibre spaces, and local homeomorphisms. Duke Math. J. **21**, 329–336 (1954)
7. Chaney, RW.: Piecewise C^k-functions in nonsmooth analysis. Nonlinear Anal. Theor. Meth. Appl. **15**, 649–660 (1990)
8. Chien, MJ., Kuh, ES.: Solving piecewise linear equations for resistive networks. Circ. Theor. Appl. **4**, 3–24 (1976)
9. Clarke, FH.: Optimization and Nonsmooth Analysis. Les publications CRM, Université de Montréal (1989)
10. Cottle, RW., Kyparisis, J., Pang, JS. (eds.): Variational inequalities. Math. Program. Ser. B **48**(2) (1990)
11. Cottle, RW., Kyparisis, J., Pang, JS. (eds.): Complementarity problems. Math. Program. Ser. B **48**(3) (1990)
12. Deimling, K.: Nonlinear Functional Analysis. Springer, New York (1985)
13. Dempe, S.: Directional differentiability of optimal solutions under Slater's condition. Math. Program. **59**, 49–69 (1993)
14. Demyanov, VF. Rubinov, AM.: Quasidifferential Calculus. Optimization Software, Inc., New York (1986)
15. Dupuis, P., Nagurney, A.: Dynamical systems and variational inequalities. Ann. Oper. Res. **44**, 9–42 (1993)
16. Eaves, BC., Rothblum, UG.: Relationships of properties of piecewise affine maps over ordered fields. Lin. Algebra Appl. **132**, 1–63 (1990)
17. Federer, H.: Geometric Measue Theory. Springer, Berlin (1969)
18. Fiacco, AV.: Introduction to Sensitivity and Stability Analysis in Nonlinear Programming. Academic, New York (1983)
19. Fletcher, R.: Practical Methods of Optimization. John Wiley, Chichster (1987)
20. Fujisawa, T., Kuh, ES.: Piecewise-linear theory of nonlinear networks. SIAM J. Appl. Math. **22**, 307–328 (1972)

21. Gauvin, J., Janin, R.: Directional behaviour of optimal solutions in nonlinear mathematical programming. Math. Oper. Res. **13**, 629–649 (1988)
22. Grünbaum, B.: Convex Polytopes. Interscience-Wiley, London (1967)
23. Hager, WW.: Lipschitz continuity for constrained processes. SIAM J. Contr. Optim. **17**, 321–338 (1979)
24. Halkin, H.: Implicit functions and optimization problems without continuous differentiability of the data. SIAM J. Contr. **12**, 229–236 (1974)
25. Harker, PT., Pang, JS.: Finite-dimensional variational inequality and nonlinear complementarity problems: a survey of theory, algorithms and applications. Math. Program. Ser. B **48**(2), 161–220 (1990)
26. Hudson, JFP.: Piecewise Linear Topology. W.A. Benjamin, Inc., New York (1969)
27. Jongen, HTh., Pallaschke, D.: On linearization and continuous selections of functions. Optimization **19**, 343–353 (1988)
28. Klatte, D., Tammer, K.: Strong stability of stationary solutions and Karush-Kuhn-Tucker points in nonlinear optimization. Ann. Oper. Res. **27**, 285–308 (1990)
29. Kojima, M., Saigal, R.: A study of PC^1-homeomorphism on subdivided polyhedrons. SIAM J. Math. Anal. **10**, 1299–1312 (1979)
30. Kojima, M., Saigal, R.: On the relationship between conditions that insure a PL mapping is a homeomorphism. Math. Oper. Res. **5**, 101–109 (1980)
31. Kojima, M.: Strongly stable stationary solutions in nonlinear programming. In: Robinson, SM. (ed.) Analysis and Computation of Fixed Points, pp. 93–138. Academic, New York (1980)
32. Kolmogorov, AN., Fomin, SV.: Introductory Real Analysis. Prentice-Hall Inc., Englewood Cliffs (1970)
33. Kuhn, D., Löwen, R.: Piecewise affine bijections of $\mathbb{R}^n$, and the equations $Sx^+ - Tx^- = y$. Lin. Algebra Appl. **96**, 109–129 (1987)
34. Kummer, B.: Newton's method for non-differentiable functions. In: Guddat J., et al. (eds.) Advances in Mathematical Optimization, pp. 114–125. Akademie-Verlag, Berlin (1988)
35. Kummer, B.: The inverse of a Lipschitz function in $\mathbb{R}^n$: complete characterization by directional derivatives. IIASA-Working paper WP-89–084 (1989)
36. Kummer, B.: Lipschitzian inverse functions, directional derivatives, and applications in $C^{1,1}$-optimization. J. Optim. Theor. Appl. **70**, 559–580 (1991)
37. Kummer, B.: Newton's method based on generalized derivatives for nonsmooth functions: convergence analysis. In: Oettli, W., Pallaschke, D. (eds.) Advances in Optimization, pp. 171–194. Springer, Berlin (1992)
38. Kuntz, L.: Topological ascpects of nonsmooth optimization, Habilitation Thesis, Institut für Statistik und Mathematische Wirtschaftstheorie, Universität Karlsruhe, 76128 Karlsruhe, Germany (1994)
39. Kuntz, L.: An implicit function theorem for directionally differentiable functions. J. Optim. Theor. Appl. **86**, 263–270 (1995)
40. Kuntz, L., Scholtes, S.: Structural analysis of nonsmooth mappings, inverse functions, and metric projections. J. Math. Anal. Appl. **188**, 346–386 (1994)
41. Kuntz, L., Scholtes, S.: Qualitative aspects of the local approximation of a piecewise differentiable function. Nonlinear Anal. Theor. Meth. Appl. **25**, 197–215 (1995)
42. Kyparisis, J.: Sensitivity analysis for nonlinear programs and variational inequalities with nonunique multipliers. Math. Oper. Res. **15**, 286–298 (1990)
43. Lemke, CE., Howson, JT.: Equilibrium points of bimatrix games. SIAM Rev. **12**, 45–78 (1964)
44. Luo, ZQ., Pang, JS., Ralph, D.: Mathematical Programs with Equilibrium Constraints. Cambridge University Press, Cambridge (1996)
45. Mangasarian, OL.: Nonlinear Programming. McGraw-Hill, New York (1969)
46. Mangasarian, OL., Solodov, MV.: Nonlinear complementarity as unconstrained and constrained minimization. Math. Program. Ser. B **62**, 277–297 (1993)
47. Murty, KG.: Linear complementarity, Linear and Nonlinear Programming. Heldermann Verlag, Berlin (1988)

48. Ohtsuki, T., Fujisawa, T., Kumagai, S.: Existence theorems and a solutions algorithm for piecewise linear resistive networks. SIAM J. Math. Anal. **8**, 69–99 (1977)
49. Ortega, JM., Rheinboldt, WC.: Iterative Solution of Nonlinear Equations in Several Variables. Academic, New York (1970)
50. Palais, RS.: Natural opterations on differential forms. Trans. Am. Math. Soc. **92**, 125–141 (1959)
51. Pallaschke, D., Recht, P., Urbański, R.: On locally Lipschitz quasidifferentiable functions in Banach-space. Optimization **17**, 287–295 (1968)
52. Pang, J.S.: A degree-theoretic approach to parametric nonsmooth equations with multivalued perturbed solution sets. Math. Program. Ser. B **62**, 359–383 (1993)
53. Pang, JS.: Newton's method for B-differentiable equations. Math. Oper. Res. **15**, 311–341 (1990)
54. Pang, JS., Ralph, D.: Piecewise smoothness, local invertibility, and parametric analysis of normal maps, manuscript, University of Melbourne, Department of Mathematics, Preprint Series No. 26–1993, (1993)
55. Ralph, D.: A proof of Robinson's homeomorphism theorem for pl-normal maps. Lin. Algebra Appl. **178**, 249–260 (1993)
56. Ralph, D.: On branching numbers of normal manifolds. Nonlinear Anal. Theor. Meth. Appl. **22**, 1041–1050 (1994)
57. Ralph, D., Dempe, S.: Directional derivatives of the solution of a paramtric nonlinear program. Math. Program. **70**, 159–172 (1995)
58. Ralph, D., Scholtes, S.: Sensitivity analysis and Newton's method for composite piecewise smooth equations. Math. Program. **76**, 593–612 (1997)
59. Rheinboldt, WC., Vandergraft, JS.: On piecewise affine mappings in $\mathbb{R}^n$. SIAM J. Appl. Math. **29**, 680–689 (1975)
60. Rice, JR.: A theory of condition. SIAM J. Numer. Anal. **3**, 287–310 (1966)
61. Robinson, SM.: An implicit function theorem for a class of nonsmooth functions. Math. Oper. Res. **16**, 292–309 (1991)
62. Robinson, SM.: Generalized equations. In: Bachem, A., Grötschel, M., Korte, B. (eds.) Mathematical Programming: The State of the Art, pp. 346–367. Springer, Berlin (1983)
63. Robinson, SM.: Homeomorphism conditions for normal maps of polyhedra. In: Ioffe, A., Marcus, M., Reich, S. (eds.) Optimization and Nonlinear Analysis, pp. 240–248. Longman, Essex (1992)
64. Robinson, SM.: Local structure of feasible sets in nonlinear programming, Part III: Stability and sensitivity. Math. Program. Study **30**, 45–66 (1987)
65. Robinson, SM.: Mathematical foundations of nonsmooth embedding methods. Math. Program. **48**, 221–229 (1990)
66. Robinson, SM.: Normal maps induced by linear transformations. Math. Oper. Res. **17**, 691–714 (1992)
67. Rockafellar RT.: Convex Analysis. Princeton University Press, Princeton (1970)
68. Rourke, CP., Sanderson, B.J.: Introduction to Piecewise-Linear Topology. Springer, New York (1982)
69. Samelson, H., Thrall, RM., Wesler,O.: A partition theorem for Euclidean space. Proc. Am. Math. Soc. **9**, 805–907 (1958)
70. Schramm, R., On piecewise linear functions and piecewise linear equations. Math. Oper. Res. **5**, 510–522 (1980)
71. Scholtes, S.: A proof of the branching number bound for normal manifolds. Lin. Algebra Appl. **246**, 83–95 (1996)
72. Scholtes, S.: Homeomorphism conditions for coherently oriented piecewise affine mappins. Math. Oper. Res. **21**, 955–978 (1996)
73. Schrijver, A.: Theory of Linear and Integer Programming. Wiley, Chichester (1986)
74. Schwartz, JT.: Nonlinear Functional Analysis. Gordon and Breach Science Publishers, New York (1969)

75. Shapiro, A.: On concepts of directional differentiability. J. Optim. Theor. Appl. **66**, 477–487 (1990)
76. Shapiro, A.: Sensitivity analysis of nonlinear programs and differentiability properties of metric projections. SIAM J. Contr. Optim. **26**, 628–645 (1988)
77. Singer, I.: Best Approximation in Normed Linear Spaces by Elements of Linear Subspaces. Springer, New York (1970)
78. Spanier, EH.: Algebraic Topology. McGraw-Hill, New York (1966)
79. Whitehead, JHC.: On C^1-complexes. Ann. Math. **41**, 809–824 (1940)
80. Womersley, RS.: Optimality conditions for piecewise smooth functions. Math. Program. Study **17**, 13–27 (1982)
81. Zarantonello, EH.: Projections on convex sets in Hilbert space and spectral theory. In: Zarantonello, EH. (ed.) Contributions to Nonlinear Functional Analysis, pp. 237–424. Academic, New York (1971)
82. Ziegler, GM.: Lectures on Polytopes. Springer, New York (1994)

Index

S. Scholtes, *Introduction to Piecewise Differentiable Equations*, SpringerBriefs
in Optimization, DOI 10.1007/978-1-4614-4340-7, © Stefan Scholtes 2012